INNOVATION AND SUSTAINABLE MANUFACTURING

Woodhead Publishing Reviews: Mechanical Engineering Series

INNOVATION AND SUSTAINABLE MANUFACTURING

RESEARCH AND DEVELOPMENT

Edited by

Carolina Machado
J. Paulo Davim

ELSEVIER

WP
WOODHEAD
PUBLISHING
An imprint of Elsevier

Woodhead Publishing is an imprint of Elsevier
50 Hampshire Street, 5th Floor, Cambridge, MA 02139, United States
The Boulevard, Langford Lane, Kidlington, OX5 1GB, United Kingdom

ISBN: 978-0-12-819513-0 (print)

ISBN: 978-0-12-819561-1 (online)

For information on all Woodhead Publishing publications
visit our website at https://www.elsevier.com/books-and-journals

Publisher: Matthew Deans
Acquisitions Editor: Brian Guerin
Editorial Project Manager: Zsereena Rose Mampusti
Production Project Manager: Sajana Devasi PK
Cover Designer: Christian J. Bilbow

Typeset by MPS Limited, Chennai, India

Contents

List of contributors

Cem Alparslan Faculty of Engineering and Architecture, Department of Mechanical Engineering, Recep Tayyip Erdogan University, Rize, Turkey

Şenol Bayraktar Faculty of Engineering and Architecture, Department of Mechanical Engineering, Recep Tayyip Erdogan University, Rize, Turkey

Carla Marisa Rebelo de Magalhães Transdisciplinary Research Center of Innovation & Entrepreneurship Ecosystems, Lusófona University, Porto, Portugal

Vasilii Erokhin Harbin Engineering University, Harbin, P.R. China

Diana Fernandes School of Economics and Management, University of Minho, Braga, Portugal

Eglantina Hysa Department of Economics, Epoka University, Tirana, Albania

Costin Ilincă Petroleum-Gas University of Ploieşti, Ploieşti, Romania

Carolina Feliciana Machado School of Economics and Management, University of Minho, Braga, Portugal

Anilkumar Malaga Department of Production Engineering, National Institute of Technology, Tiruchirappalli, Tamil Nadu, India

Mirela Panait Petroleum-Gas University of Ploieşti, Ploieşti, Romania; Institute of National Economy, Bucharest, Romania

M. Estela Peralta Department of Design Engineering, Polytechnic School, University of Seville, Seville, Spain

Marius Gabriel Petrescu Petroleum-Gas University of Ploieşti, Ploieşti, Romania

Lukman Raimi Universiti Brunei Darussalam (UBD), Brunei

Gao Tianming Harbin Engineering University, Harbin, P.R. China

Maria Tănase Petroleum-Gas University of Ploieşti, Ploieşti, Romania

S. Vinodh Department of Production Engineering, National Institute of Technology, Tiruchirappalli, Tamil Nadu, India

About the editors

Carolina Feliciana Machado received her Ph.D. degree in Management Sciences (Organizational and Policies Management/Human Resources Management) from the University of Minho in 1999, Master's degree in Management (Strategic Human Resource Management) from Technical University of Lisbon in 1994, and Degree in Business Administration from the University of Minho in 1989. Teaching in the area of Human Resources Management since 1989 at the University of Minho, she has been an associate professor since 2004 with experience and research interest areas in the fields of Human Resource Management, International Human Resource Management, Human Resource Management in SMEs, Training and Development, Emotional Intelligence, Management Change, Knowledge Management and Management/HRM in the Digital Age, and Business Analytics. She is the head of the Human Resources Management Work Group at the School of Economics and Management at University of Minho, coordinator of Advanced Training Courses at the Interdisciplinary Centre of Social Sciences, a member of the Interdisciplinary Centre of Social Sciences at the University of Minho, and the chief editor of the International Journal of Applied Management Sciences and Engineering. She has been a guest editor of journals, book editor, and book series editor as well as a reviewer for various prestigious international journals. She has also published, both as editor/coeditor and as author/coauthor, several books, book chapters, and articles in journals and conferences.

J. Paulo Davim is a full professor at the University of Aveiro, Portugal. He is also an honorary professor at several universities, colleges, and institutes in China, India, and Spain. He received his Ph.D. degree in Mechanical Engineering in 1997, M.Sc. degree in Mechanical Engineering (Materials and Manufacturing Processes) in 1991, Mechanical Engineering degree (5 years) in 1986 from the University of Porto, the Aggregate title (Full Habilitation) from the University of Coimbra in 2005, and D.Sc. (Higher Doctorate) from London Metropolitan University in 2013. He is a senior chartered engineer of the Portuguese Institution of Engineers with an MBA and Specialist titles in Engineering and Industrial Management as well as in

Metrology. He has been designated Eur Ing by the FEANI-Brussels and a Fellow of IET-London. He has more than 35 years of teaching and research experience in Manufacturing, Materials, and Mechanical and Industrial Engineering, with special emphasis in Machining and Tribology. He also has interests in Management, Engineering Education, and Higher Education for Sustainability. He has guided large numbers of postdoctorate, Ph.D., and Master's students and has coordinated and participated in several financed research projects. He has received several scientific awards and honors. He has worked as an evaluator of projects for the European Research Council and other international research agencies and as an examiner of Ph.D. thesis for many universities in different countries. He has been the editor-in-chief of several international journals, guest editor of journals, book editor, book series editor, and scientific advisor for many international journals and conferences. He is currently an editorial board member of 30 international journals and acts as a reviewer for more than 100 prestigious Web of Science journals. He has published, as editor (and coeditor) more than 250 books and, as author or coauthor, more than 15 books, 100 book chapters, and 600 articles in journals and conferences (more than 300 articles in journals indexed in Web of Science core collection/h-index 63 + /13500 + citations, SCOPUS/h-index 68 + /17000 + citations, and Google Scholar/h 63 + /-index 88 + /27500 + citations). He has been listed in World's Top 2% Scientists by Stanford University study.

Preface

This book is intended to cover the issues related to innovation and sustainable manufacturing in a context in which organizations are facing, day after day, challenges related to the continuous changes in the markets as well as in the environment as a whole.

In the manufacturing environment, innovation and sustainability need to be present in the different ways to manage, think about, and understand today's organizations in order for them to become even more competitive and act strategically in the markets where they are and want to be present in the near future. Indeed, innovation is more and more crucial to the continuing success and survival of any organization. The term *sustainable manufacturing* refers to the process of manufacturing that is concerned not only with present, but also with the future. In other words, it refers to the new ways of production that are in line with the needs of the present without compromising the ability of future generations to meet their own needs. In an organizational environment, when we consider innovation and sustainability, we are creating the support that will enable companies to create products, services, and processes that are good for both society and the organization. Focusing particularly on sustainable manufacturing, organizations look to manufacture new and innovative products through economically sound processes that minimize negative environmental, economic, and social impacts while they are concerned with the needs and interests of future generations.

This book, organized in eight chapters, can be used by academics, researchers, human resources managers, managers, engineers, and other professionals in related fields.

Following its main aims, this book covers the field of *Innovation and Sustainable Manufacturing* in eight chapters: Chapter 1 discusses "Sustainable and fractal manufacturing systems: complexity minimization when sustainability and Industry 4.0 interact"; Chapter 2 focuses "Reliability of manufacturing technologies"; Chapter 3 contains information about "Energy innovations and sustainable development of circumpolar territories in Russia," while Chapter 4 deals with the "Analysis of factors influencing smart and sustainable manufacturing systems using a multicriteria decision-making tool." Chapter 5 covers "Catching up with sustainable development in emerging markets through financial innovation"; while Chapter 6 speaks about "Corporate social responsibility, business ethics, sustainability, and innovation: a theoretical analysis of the importance of these concepts in times of pandemic." Chapter 7 discusses "Sustainability: leadership and reporting as its pillars," and finally, Chapter 8 covers "Comparison of the SLM, SLS, and DLMS techniques in additive manufactured of AlSi10Mg alloys."

Innovation and Sustainable Manufacturing can serve as a critical tool for academics, researchers, human resource managers, managers, engineers, and other professionals. The interest in this subject is evident for many types of organizations all over the world, in particular different institutes and universities.

Carolina Machado
J. Paulo Davim

Acknowledgments

The editors acknowledge their gratitude to Elsevier for this opportunity and for their professional support. Finally, we would like to thank all the chapter authors for their interest and availability to work on this project.

CHAPTER

1

Sustainable and fractal manufacturing systems: complexity minimization when sustainability and Industry 4.0 interact

M. Estela Peralta

Department of Design Engineering, Polytechnic School, University of Seville, Seville, Spain

1.1 Introduction

The unpredictable dynamics of the market, with needs that must be met in short periods of time and increasingly stringent demands in terms of quality and variety of products, has led to a considerable increase in complexity in manufacturing systems. In this situation, cooperation between organizations (formation of self-organized clusters) is essential to maintain market shares and to ensure their survival with rapid adaptation to changes. New trends have emerged for the organizations of the 21st century, in which the concepts of digitalization, intelligence, and real-virtual duality are relevant to manage complexity. These will be the main indicators that the projects involving research, development, and innovation will use to organize themselves to contribute to the development of the Fourth Industrial Revolution, sometimes known as Industry 4.0.

The concept of Industry 4.0 is a new approach that is advancing rapidly in the industrial sector. It promotes the transformation of organizations toward new conceptions of connected, intelligent, and sustainable manufacturing systems (IDE, 2016; Sauter, Bode, & Kittelberger, 2015).

Innovation and Sustainable Manufacturing
DOI: https://doi.org/10.1016/B978-0-12-819513-0.00008-1

Sustainability is one of the pressing lines of research in Industry 4.0. It encompasses three dimensions of study: economic (which includes the technical feasibility of the results), environmental (which considers the efficiency of the process in terms of minimal impact on the ecosystem with which the manufacturing system interacts), and social (with a distinct interest in manufacturing that focuses on the user as well as promoting the safety, health, and comfort of the workers).

The development of these areas is motivated by the results of the production activities of the past. In particular, the manufacturing industry has contributed to the increase in the global impact by wasting energy and material resources; causing environmental deterioration, destruction of ecosystems, and loss of biodiversity; and, for certain sectors of the population, a reduction in health, well-being, and quality of life (Peralta, Marcos, & Aguayo, 2016). The new tendency is to develop strategies based on excellence, continuous adaptability, and openness to innovation, oriented by cyclicality, efficiency and ecocompatibility with the ecosystem and simultaneously, focused on achieving solutions with a balance among economic, environmental, and social results. A sustainable manufacturing system with these characteristics introduces a large volume of information, data, requirements, tasks, and new control techniques that need to be considered in the design, configuration, and procedures of the manufacturing process. For these solutions to advance, new frameworks are required that include theoretical foundations, models, and experimental research, whose scope integrates (1) how to manage the complexity that arises in the interaction between conventional and sustainable requirements of the manufacturing system and (2) the search for the tradeoff among the three sustainable dimensions (economy, ecology, and equity).

To carry it out, this research extends the scope of the fractal theory by adapting it to sustainable manufacturing engineering and it proposes a framework to manage complexity by defining the sustainable and fractal manufacturing systems (SF|Ms). In the next section a review of the literature on the evolution of complexity management in manufacturing systems is presented and discussed. In addition, the interaction between sustainability and manufacturing is analyzed, identifying which lines of research are necessary for the development of possible theoretical foundations, models or experimental research. Then the fundamentals and main advantages of the fractal theory applied to manufacturing are presented. A framework is then proposed for SF|Ms as a strategy to reduce complexity and to find balanced sustainability in manufacturing processes. Finally, conclusions and future development are presented.

1.2 Complexity in sustainable manufacturing systems

In manufacturing, the management of complexity models has evolved according to the requirements that need to be controlled in the production process (volume of production, quality, flexibility, adaptability, efficiency, etc.) until achieving the present approaches of chaos theory, complex adaptive systems (CASs), and resilient systems. To understand the interaction that exists between sustainability and manufacturing, one needs to analyze the evolution of the different models that will be the starting point to establish the theoretical basis and the new frameworks for sustainable manufacturing systems.

1.2.1 The evolution of complexity management in manufacturing systems

Complexity is a parameter that is of interest in the design, management, and control of systems during their life cycle. Studies on this topic have been carried out in a variety of areas of expertise, including marketing, economic systems, engineering, social groupings, even natural systems, for human cognition, atmospheric phenomena, and ecosystems.

The term *complexity* has evolved according to the way of addressing the behavior of artificial systems (Simon, 2006). Simon divides the evolution of complexity in three stages. The first stage coincided with World War I; the concept of holistic systems arose, moving away from the reductionist idea of those times in which a system was identified as the sum of all its parts. The second stage, during World War II, began with the birth of the cyber-age, with the development of the theory of information, the theory of servomechanism (or feedback control systems), and the possibility of storing data on computers. Complexity was explained from the point of view of the reduction of entropy (disorder) and homeostasis or self-stabilizing of the system. Technological advances enabled observation, with feedback, of how a system could orient its activity to achieve its objectives by adapting to its environment or changing surroundings. In this second stage, complex systems were equipped with the abilities to (1) orient themselves toward the function for which they were developed (self-organization), (2) discover and define its task, (3) self-analyze and detect its execution as well as the discordance with the marked objective, and (4) solve, via a set of actions, the discrepancies that hinder its efficient performance (dynamism and self-optimization). Some strategies from this second stage remain; these involve complexity, focusing on abstracting the specific properties from the different complex systems (social, biological, and physical) to look for similarities and understand or transpose natural behaviors to artificial ones (analogy).

In the third stage (present day), although it is closely related to the second, the interest focuses on the need to understand the functioning of the systems at a global level (e.g., the environment, the market, or society). The concepts of chaos and CASs are used (Weichhart, 2013). The investigation centers its study on managing mechanisms and controlling complexity. The proposed models, methodologies, and techniques are included for analysis; in addition, new tools have been developed to study the behavior of the systems, such as that of artificial intelligence and metaheuristics, cellular automata, chaos theory, or fractal systems.

Chaos theory is one of the most important contributions that have been made to the topic of complexity. Chaotic systems are deterministic dynamical systems that part from the initial conditions that are given. A small modification or disturbance in them may cause large deviations in their intended behavior. They are characterized by unpredictable and random behavior (although their elements are regulated by laws and principles) and are highly sensitive to their initial conditions. Since Poincaré's studies in 1946 (Poincare, 1946), chaos theory has developed slowly. However, it has become established an important conceptual framework, providing methodologies and tools to assess certain types of systems and substantiate their behavior (Barkman, 1989).

In manufacturing systems, chaos theory is being used in production management as an approach that integrates the control of both deterministic situations and those that are nondeterministic. There are some interesting and untapped publications that analyze and propose models and measures of complexity in manufacturing systems (De Toni, Nardini, Nonino, & Zanutto, 2001; Deshmukh, Talavage, & Barash, 1998; Efthymiou, Pagoropoulos, Papakostas, Mourtzis, & Chryssolouris, 2012; ElMaraghy & Urbanic, 2004; ElMaraghy, ElMaraghy, Tomiyama, & Monostori, 2012; Frizelle & Woodcock, 1995; Kumar, 1987; Mattsson, Gullander, & Davidsson, n.d.; Papakostas, Efthymiou, Mourtzis, & Chryssolouris, 2009; Vega, 2012). Of the previous two, the deterministic approaches are the most accepted, although it is not yet easy or possible to acknowledge the causes and consequences in studying cause-effect relationships. They are dealt with taking into account that a cause-effect relationship exists in operations; and based on historical data and statistical data taken from the results, the variables of any manufacturing process are predicted and controlled. On the other hand, there are certain situations for which manufacturing systems cannot be considered deterministic, including the impacts (sustainability), security, reliability, the human factor, quality, the demand for products and services, changes in customer preferences, planning and execution of production, the evolution of the market, and the behavior of the competition. These situations need to take chaotic behaviors into consideration.

1.2.2 The integration of deterministic and chaotic systems to manage complexity in manufacturing

Chaos theory was first applied to manufacturing systems in 1991 by Moody and Morley (1999) and Morley (1996), who studied and simulated paint booths that worked as autonomous agents, selecting their tasks in the manufacturing process of the bodywork of trucks for General Motors; since its proposal, scientists, researchers, and engineers have been studying deterministic chaotic behaviors in order to develop new tools that can identify the unpredictable, stochastic, or irregular behavioral patterns that are detected in manufacturing processes, that is, to manage the complexity that is linked to such systems by predicting its behavior.

The Santa Fe Institute is currently the place of reference in studying chaos theory and complexity theory; it has developed the fundamental principles of CASs in the fields of physics, biology, and social or computational systems. Authors and researchers have evaluated the chaos of manufacturing systems. Beaumariage and Kempf (1994), Deshmukh (1993), and McCarthy, Rakotobe-Joel, and Frizelle (2000) described the chaotic nature of the manufacturing systems. More specifically, Horns (1989) studied control of the manufacturing that is oriented toward processes (job shop); Alfaro and Sepulveda (2006) focused on the chaotic behavior of a production reactive system in which there was no previously determined schedule and the operations were assigned to the machines according to the system's condition; Chyssolouris, Giannelos, Papakostas, and Mourtzis (2004) concentrated on production planning; Karaçali, Halil, and Demirci (2009) examined the information systems in manufacturing companies; and Huaccho Huatuco, Efstathiou, Sivadasan, and Calinescu (2001) analyzed the dynamic complexity in computer-aided manufacturing (CAM) by researching metrics to then evaluate and measure, parting from entropy.

From previous studies, one can conclude that in general, manufacturing systems will occasionally require a deterministic approach, in particular, to manage material flows, energy, and information (Attar & Kulkarni, 2014) by using exact methods based on logical rules, which will be used to model and optimize many of the system's functions. With regard to the knowledge and management of other areas of the production process, a deterministic dynamic will not usually suffice. Therefore these solutions are oriented toward dynamic processes that have a high uncertainty and in which unexpected events and situations occur very often and for which it will be necessary to use complexity management techniques. Some examples are control of the market with the introduction of innovations in products, management of the products' life cycle in the market, the use of industrial organizational techniques, the integration of new technologies into the production processes, the results from the manufacturing operations (quality), the behavior of human resources and stakeholders, and the evolution of the market and the customer; all are required in order to achieve balanced results sought out for sustainable production systems.

1.2.3 Conclusions to manage complexity in sustainable production systems

Both perspectives, deterministic and chaotic, are integrated in the conventional functioning of a production system. Based on this idea, through a set of experiments, Beaumariage and Kempf (1994) established three main conclusions to guide the complexity management process:

- Working with complex systems in an optimal manner may require finding a balance or equilibrium, a boundary between a deterministic and chaotic perspective.
- A small change in some parameters could provoke the system to transition to a chaotic behavior, reducing the system's performance.
- Attempts to make a system work tolerantly toward this limit in order to prevent the transition to a chaotic mode could reduce the results' performance.

With regard to how to orient management models, Gálvez-Medina, Balankin, and Resenos (2009) and Morgan (1996) systematically described the characteristics of complexity in manufacturing systems:

- Complexity is an information measure of the system that is used to characterize its function and structure and the interconnections of the elements of its architecture.
- The complexity of a system depends on the number and variety of elements and their interactions. In addition, the interaction of the elements is the origin of a behavior that is not present in the individual elements in the system (emergence).
- Complexity is an evolutionary process that involves the nonlinear behavior that results from successive crises (industrial revolutions, technological transitions, productive obsolescence, etc.).
- The traits that characterize a manufacturing system define it as a CAS that is motivated by an objective (goal), for which its elements are specialized, working toward continuous improvement.

It is of interest to address the complexity of manufacturing systems and see it as an opportunity to understand their behavior and efficiently manage the results. By managing the system's operation, it should be possible to identify its different conditions as well as the variety of situations in which it can guarantee the fulfillment of the objective according to the established requirements. This will be possible with the unfolding of the types and dimensions of complexity (structural, technical, functional, directional, etc.), all of which are aspects that modify the condition and behavior of the system but that should be treated simultaneously for maximum performance and improved balanced results (see Section 1.4.5).

One of the possible solutions in finding the stochastic-deterministic balance described is by addressing the design and development of organizations and the manufacturing systems from the CAS foundations. This will enable the integration of all the functioning levels of the manufacturing system as a combination of ordered and chaotic conditions because the CAS has an architecture with a multidimensional function. On the one hand, the procedure of the deterministic system controls, in an orderly manner and through a set of constraints and premeditated action rules, the behavior of the systems' elements. On the other hand, the dynamics of the chaotic system allows the elements to be autonomous, defining (1) their own forms of conduct, (2) their operating limits, and (3) the rules in those situations that require it. Autonomy (in the sense of independence) aims to increase the flexibility of a reaction to sudden changes in the context for which no preconfigured protocols exist (in which there is no external influence) and for which the agents have the ability and initiative to solve the resulting problems and redirect their objective. In the case of the organization or the manufacturing system it is carried out with innovation processes.

A system that is in constant equilibrium (without continuous improvement) does not have the ability to respond to its environment and therefore cannot adapt quickly; by contrast, a completely chaotic system will cease to operate as a system due to its own behavior. The most productive condition (which characterizes the CAS systems) takes place at the edge of chaos, where there will be maximum variety and creativity and innovation will enable the development of new possibilities in order to survive (Fryer & Ruis, 2004). In this sense, the union of the two perspectives enables coevolution in a synergistic manner: The system (at a global or macroscopic level) limits the behavior of the elements; at the same time, the elements (at an individual or microscopic level) autonomously modify the restrictions of the global system, turning the system into a self-organizing one.

To configure this type of hybrid system, there are some frameworks that are applied in the field of manufacturing and contribute a set of principles and strategies for its design and usage as well as including complexity management procedures. The ones of interest are those that are applied from a classical perspective of intelligent and distributed manufacturing: the bionic, holonic, and fractal approaches (Chituc & Restivo, 2009; Kühnle, 2009). Additionally, its knowledge base centered on bionic processes (based on natural processes) makes them suitable to introduce the challenges that the need for sustainability places on manufacturing systems. All three include agile manufacturing as an evolution from lean manufacturing (Tharumarajah, 1996; Tharumarajah, Wells, & Nemes, 1998). Its benefits include delivering the highest quality with minimal resources, controlling overproduction, and reducing waiting time, transportation, excess processing, inventory, movements, and defaults. Agile manufacturing also contributes to increasing the flexibility of production and fast customer response.

Although they have the same objective, bionic, holonic, and fractal manufacturing provide different benefits. They differ mainly in the following aspects: (1) level of autonomy of the system's entities, (2) mechanisms of self-regulation, and (3) adaptation to changes in the context (Chituc & Restivo, 2009; Kühnle, 2009; Kwangyeol, 2003; Silva & Ramos, 1999; Tharumarajah et al., 1998; Tharumarajah, 1996).

This work considers fractal theory to be the most appropriate of the three to manage complexity in sustainable manufacturing systems. This decision is supported by the studies of Warnecke (1993), who proposes managing the two conditions, deterministic and chaotic, simultaneously in the organizations and in the manufacturing systems from the fractal properties. In addition, the research of Peralta, Marcos, Aguayo, and Lama (2017), Peralta, Marcos, Aguayo, Lama and Córdoba (2015), and Peralta-Alvarez (2016) specifies and defines the benefits for the fractal paradigm by managing complexity when manufacturing and sustainability interact, thus justifying how the fractal paradigm is an opportunity to incorporate search mechanisms and control of the requisite variety (Ashby, 1991) in sustainable production systems. The synthesized reasons supporting this decision are the following:

- The fractal paradigm includes a set of principles that enable the creation of CASs as an optimal way to combine the characteristics of deterministic and chaotic systems, thus creating self-organized systems.
- A fractal organization will act as the interface to find the condition of quasiequilibrium between the chaotic and deterministic behaviors within the state space, using self-similar structures of control.
- A fractal configuration creates systems with a deterministic chaos behavior. Its temporary evolution will apparently be irregular but will be far from an entropic condition. In other words, it will be deterministic as long as the initial conditions of the system do not change.
- The fractal paradigm is a framework in keeping with smart manufacturing systems and that Industry 4.0 proposes. It can add value to products with minimum variety and complexity management.

- The fractal paradigm is a flexible framework that can adapt to the design, configuration, and use of sustainable and smart manufacturing systems that are tailored to the requirements of Industry 4.0: self-organized, efficiency, self-similarity, self-optimization, and resilience.

The results of this research have given rise to a framework of knowledge that (1) expands the scope of the fractal paradigm, adapting it to the new requirements of complexity management when the sustainable and conventional dimensions of the manufacturing systems interact, and (2) proposes the SF|Ms in order to structure new architectures of smart and sustainable production processes in Industry 4.0.

1.3 Fractal paradigm: foundations

The fractal paradigm is a scientific discipline and knowledge base that is intended to aid in the design and development of systems that are similar to a fractal entity (formal and natural) from a structural, functional and informational point of view, with them responding, adapting and surviving in a turbulent and changing environment in an autonomous way.

The fractal concept applied to industrial organization and in particular to manufacturing systems arose with mathematics and chaos theory, although it was developed to manage the inflexibility and rigidity of organizations (Warnecke, 1993). In the 1990s the first publications appeared that applied the fractal paradigm to industrial systems. The fractal paradigm was an innovative perspective that was different from the conventional concept of a business as a hierarchical structure, taking into account the characteristics of the manufacturing system similar to that of CASs (Weichhart, 2013).

Since the first publications by Warnecke (1993), Dyke and Irish (1985), and Tirpak, Daniel, LaLonde, and Davis (1992), fractal theory has been included in manufacturing engineering to solve CASs in their different conceptions (CAS, collective adaptive system, complex adaptive emergence system, complex adaptive manufacturing systems, and adaptive complex enterprise) (Anne, Bhagavatula, Chedjou, & Kyamakya, 2008; Brandt, Hartmann, Sander, & Strina, 1999; Dobrescu & Florea, 2013; Kühnle, 2009; Kwangyeol, 2003; Markfort, Martinetz, Klostermeyer, & Brehmer, 2002; Ryu & Jung, 2004; Shin, Mun, & Jung, 2009; Shin, Ryu, & Jung, 2012; Strauss & Hummel, 1995; Tirpak et al., 1992; Wiendahl & Harms, 2004; Xu & Chen, 2010). The progress and evolution of the fractal paradigm are continuing for the industry of the future by way of new research and applications, identifying the latest contributions in publications related to Industry 4.0 (Bauernhansl, Ten Hompel, & Vogel-Heuser, 2014).

In its conventional approach (economic dimension), fractal systems arose as the solution to the complexity of structuring and management of smart manufacturing, which generally included large amounts of information and a large number of variables and requirements. Its aim was to obtain greater efficiency in including value to the products with minimum variety and complexity management. This perspective enables the fractal paradigm to become an opportunity to design and configure the standardized architectures of Industry 4.0; in addition, as discussed in previous sections, it is the best suited to manage the complexity that arises from the interaction between sustainability and manufacturing.

With regard to the design process, the fractal manufacturing framework can develop manufacturing processes similar to the fractal shapes; by adjusting the granularity of its functions, requirements, and operations, the mechanisms needed to deal with the pressure of the context in which they operate are included in the processes (they adapt dynamically to changes in the environment) (Aguayo-González, Marcos-Bárcena, Sánchez-Carrilero, & Lama-Ruiz, 2007; Baldomero, Salmerón, & Aguayo, 2006). The innovative aspect of the fractal organization is the transformation of the conventional hierarchical or centralized structures into nesting structures, which provide flexibility. Distributed architectures are created in which each element is self-similar in shape and action (process, work equipment, information packages, etc.) but the elements work independently to meet the overall objective (Cha, Shin, Ryu, & Jung, 2007).

To design an industrial system that complies with these characteristics, four principles are defined, which derive from the analogy developed by Warnecke: (1) self-similarity, (2) self-organization, (3) goal orientation, and (4) dynamism. These characteristics enable the incorporation of a set of scalable requirements (instantiable) at any level or condition of the system and give them the ability to adapt and survive in the ever-changing and dynamic market. These principles have been established by analyzing the concepts and foundations of the mathematical fractals and from the principles that govern natural objects with fractal solutions in nature.

1.4 Sustainable and fractal manufacturing systems to manage complexity in Industry 4.0

In this section the framework that is proposed for sustainable fractal manufacturing is defined as a strategy to reduce complexity and to find balanced sustainability in manufacturing processes. It includes the principles of synthesis and an architecture to design SF | Ms. The proposal is guided from the current Industry 4.0, IMS2020, and Factories of the Future research projects. It includes a technological platform to configure the cyber-physical system (CPS) (Asare, Broman, Lee, Torngren, & Sunder, 2016; Jiang, Ding, & Leng, 2016; Lee, Bagheri, & Kao, 2015) that is formed by a multiagent fractal architecture to monitor and self-regulate in real time (Peralta-Alvarez, 2016).

1.4.1 The concept of sustainable and fractal manufacturing systems

A SF | Ms is a manufacturing system that acts as second nature (similar to a natural ecosystem); it operates in the technosphere (industrial, rural, or urban ecosystem). It is a system with a fractal structure and behavior (Kwangyeol, 2003; Warnecke, 1993) that is able to configure, maintain, and reproduce itself (i.e., it is self-organized and self-optimized). Furthermore, it is a system that carries out its objective without negative interference (damage or impact) with other systems by performing collaborative activities to integrate and adapt itself efficiently in its context of action (it is dynamic and goal-oriented).

1.4.2 Basic principles for sustainable and fractal manufacturing systems

The framework (Peralta et al., 2017; Peralta-Alvarez, 2016) involves three principles that establish the basis for the design, development, and evaluation procedure of a SF | Ms. Each principle extends the scope of Warnecke's fractal theory (Kwangyeol, 2003; Ryu, Son, & Jung, 2003; Warnecke, 1993), introducing as the main dimension the sustainable and balanced development of the manufacturing system; it also establishes the management procedure for complexity.

1. *PI: Principle of scale.* This principle includes the properties that are required to manage complexity through the self-similarity of the system, such as (1) product (structure, the static dimension) and (2) process (behavior, the dynamic dimension). The structure and behavior (process and internal and external interaction) of the SF | Ms are established in a manner similar to those of a basic fractal unit (BFU) (Tirpak et al., 1992), adapting to its activity. By recursively using a pattern of structure and behavior at the product and process levels, it (1) reduces complexity, (2) replicates the system (in growth, development, and evolution conditions), and (3) promotes continuous improvement by analyzing its behavior (economic, environmental, and social impact). All of this is carried out in a

similar way at any level of observation, design, management, or operation, thanks to the scale invariance property of the fractal structure. This principle enables the design and configuration of SF | Ms as similar structures by recursion (see Fig. 1.1). It facilitates the incorporation of the three dimensions of sustainability with the same procedure to ensure self-regulation and adaptation to the activity context. In the procedure of complexity management it proposes the use of the metric *self-similarity* to control this principle, with the indicator *H(s)*, which reflects the result of the scope of self-similarity between fractal patterns.

2. *PII: Principle of requisite fractality.* This principle creates the necessary technodiversity (variety and relative abundance) to adapt to its context throughout the life cycle of the fractal. The SF | Ms is an autonomous entity situated in an ecosystem; it must establish itself from the set of requirements and limits imposed by the context, its internal characteristics for survival, and minimum impact, and so it adapts its BFU continuously. This principle enables the system to adapt to the environment: SF | Ms, as open and innovative organizations, evolve in a stable manner, thanks to the regulatory mechanisms that configure their structure and behavior (self-organization). This principle includes the skills and knowledge that

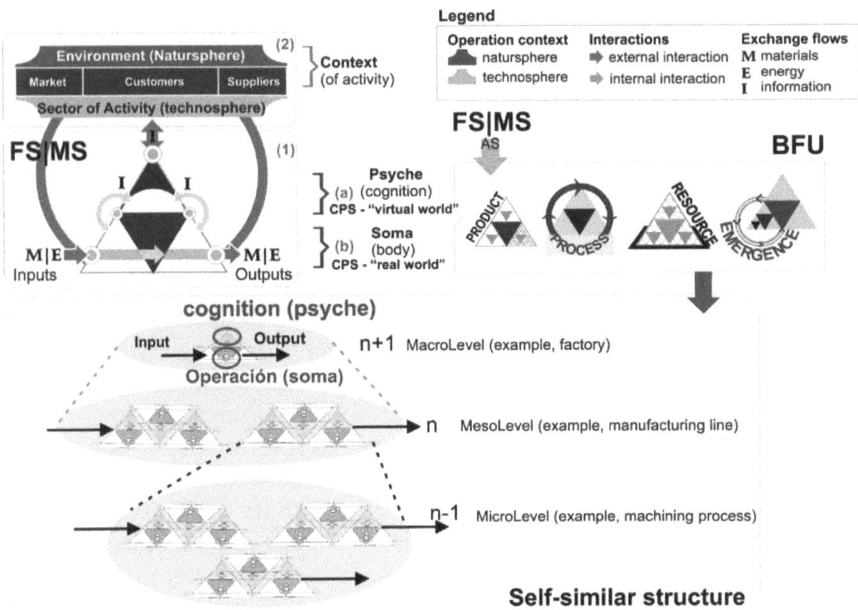

FIGURE 1.1 Fractal architecture for an sustainable and fractal manufacturing systems.

are required to identify the new conditions of smart action (active and proactive) that enable continuous improvement by modifying the internal variety of the fractals according to the short- and long-term needs (analysis in real time and prospective evolution, which is a dynamic principle). In the procedure of complexity management the use of the metric *coevolution* is proposed in order to control this principle, with the indicator *H(d)*, reflecting the evolution of the system in its behavior and in the planned and successfully executed goals.

3. *PIII: Principle of ecocompatibility*. This principle contributes emergence properties (self-organization and self-optimization) to the system in its interaction with the environment so as to be (1) ecoefficient (minimization of use of resources) and (2) ecoeffective (value generation or damage recovery). It is the synthesis of sustainability, which will be integrated as a main function of the SF | Ms. This principle establishes the adaptation of the SF | Ms to its operating context: (1) the technosphere (or industrial, urban, or rural ecosystem) and (2) the natursphere (or natural ecosystem); enabling it to behave according to the goal orientation principle as second nature. The organization of the fractal manufacturing processes from the principle of ecocompatibility will establish the architecture of the system with balanced multidimensional strategies, that is economic, environmental, and social strategies. This is done thanks to the research, development, and innovation activities of the organization according to the demands of the environment. In the procedure of complexity management the use of the metric *triple-E performance* is proposed in order to control this principle, with the C_MS_{TPh} indicator, which reflects the result of the sustainable performance (economic, environmental, and social) in the execution of its activity.

1.4.3 Generic architecture of a sustainable and fractal manufacturing systems

The design of the architecture of a SF | Ms is carried out by the BFU or self-similar pattern. The BFU is formed by four dimensions (principle I): SF | Ms = {Product, Process, Resource, Emergence}. In their interaction as a system, they will produce two processes of control and action (principle II): SF | Ms = {P + S}. They are defined as follows:

$$(SF|FS)^{virtual}_{physical} = \left[\sum_{i=1}^{n}(PrD_i + PrC_i + R_i + E_i)\right]_1 U(Px\ S)_2 U(\text{Env} + \text{Mrk} + \text{Spp} + \text{Cu})_3 \tag{1.1}$$

1. *The BFU* {PrD, PrC, R, E} is structured by the following dimensional elements:
- PrD, *Product or structure* (examination of the shape). In this dimension the system is defined from the point of view of its structure and morphology, including elements that compose it, location, and internal and external relationships.
- PrC, *Process or performance* (examination of the function). This dimension defines the behavior that is linked to the functions of operation and regulation, including homeostasis, resilience, and metabolic adjustment (industrial metabolism).
- R, *Resource or support* (examination of the activity and operation context). This dimension includes the support requirements (auxiliary materials, facilities, location, etc.) that are required by the system to carry out its function as a process and maintain its stable structure as a product.
- E, *Emergence or levels of organization* (examination of the elements in operation). This dimension, taking into account its integration into the structure of the system and the evolution of its behavior, defines the properties of the fractal as a complex system. They are the result of the cooperation between the elemental units, both horizontally and vertically, in the different levels of the organization.

The combination of the previous dimensions as a system creates the behavior or the mechanisms of action and control of the SF | Ms, which, when in interaction with the environment (activity context or support unit), define the final standard architecture of the fractal manufacturing system:

2. *Unit of behavior* {P + S}, which is structured by the following mechanisms of action and control:
   a. *P, Psyche*: This is the structural part of the system that is responsible for the metabolism of the fractal based on the exchange of information according to its specialization. It carries out all the processes and phenomena of the fractal in relation to the collection, analysis, processing, inference, and exchange of data, including decision making and the learning process. It integrates its skills and processes in the form of knowledge (cognitive processes).
   b. *S, Soma*: This is the structural part of the system that is responsible for the metabolism of the fractal based on the technosphere (exchange of matter and energy); it includes all the processes and phenomena of the fractal in relation to the exchange, processing, transformation, assimilation, and liberation of matter and energy necessary to carry out its functions (goals). It integrates the processes as a set of physical actions.

3. *Context unit* {Environment, Market, Suppliers, Customer}: The fractal executes its activity in a context that integrates the operational support with which it exchanges matter, energy, and information. It is divided into two subsystems:
   a. *Environment (natursphere)*: This is the part of the natural ecosystem that supports and locates the SF | Ms, providing among other things, raw materials. Furthermore, it is the primary recipient of the environmental impact that is generated by the SF | Ms
   b. *Context of activity*: This is the technosphere or artificial ecosystem, which may be industrial, urban, or rural. It integrates the resources that are needed to carry out the activity of the fractal and is made up of organizations that exchange goods and services (metabolism):, that is, the market (competitors), suppliers, clients, and the economic-legal system.

The BFU is used recursively to form the structure of the SF | Ms. It contains all the information to create and manage the dimensional elements of the SF | Ms (product, process, resources, and emergence). Its content must be instantiated (adapted) to the aspects of complexity that have been selected and to the level of operation (process, manufacturing line, factory, enterprise, industrial park, etc.). Figs. 1.2 and 1.3 explain the data structure and information content of the BFU.

The BFU is the pattern that is used in the construction and generation of the architecture of the sustainable fractal system. Thus, although the SF | Ms includes the requisite variety (Ashby, 1991) to execute its objective and function, the structure and behavior shall comply with the condition of invariance scale; that is, at any scale level, this BFU can be easily recognized so as to analyze, modify, and manage it. For example,

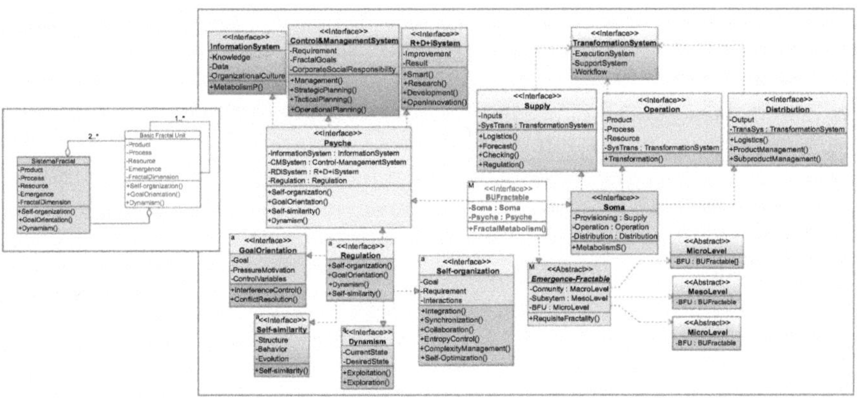

FIGURE 1.2 Basic fractal unit.

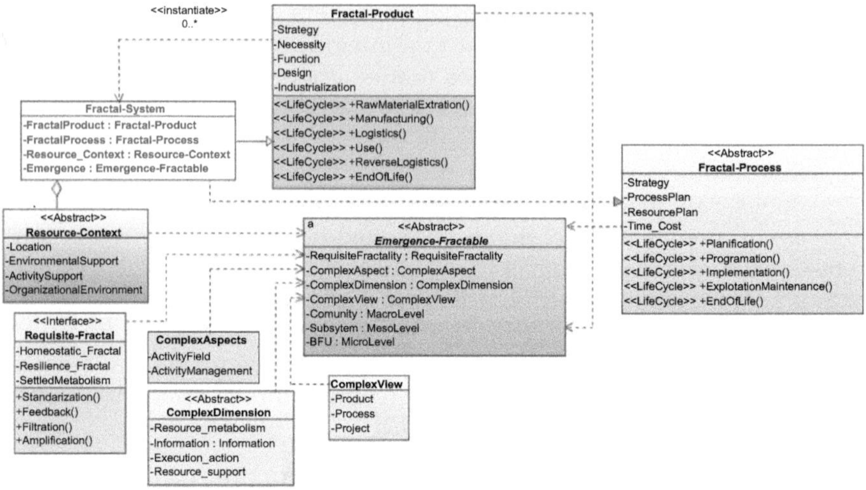

FIGURE 1.3 Data structure of the dimensional elements in sustainable and fractal manufacturing systems.

in a manufacturing system, all the levels of (1) manufacturing process, (2) manufacturing line, (3) manufacturing plant, and (4) organization and company will have the same physical structure (real), information (virtual), and behavior as a CPS.

1.4.4 Coordination and control of the metabolism (material, energy, and information) to manage the sustainability of sustainable and fractal manufacturing systems

The fractals are coordinated and controlled at scale level (multiscale) and hierarchical level (multilevel) by the principles of self-management (Kwangyeol, 2003; Ryu & Jung, 2003; Warnecke, 1993); these will be carried out by the "psyche" of the fractal: self-organization, self-similarity, goal orientation, and dynamism. This coordination is necessary because of the level of autonomy and independence that each fractal unit has within the SF|Ms; coordination is necessary to prevent interference between objectives and to manage decision making and results. The two elements, body (soma) and cognition (psyche), are connected via a communication system whose objective is the coordination and control of the fractal (matter, energy, and information) exchanging specific information packages (see Fig. 1.4).

By combining the elements P and S and by using the formation processes of fractals (processes of growth), expansive or contractive construction, and specialization (Kwangyeol, 2003), it will be possible to

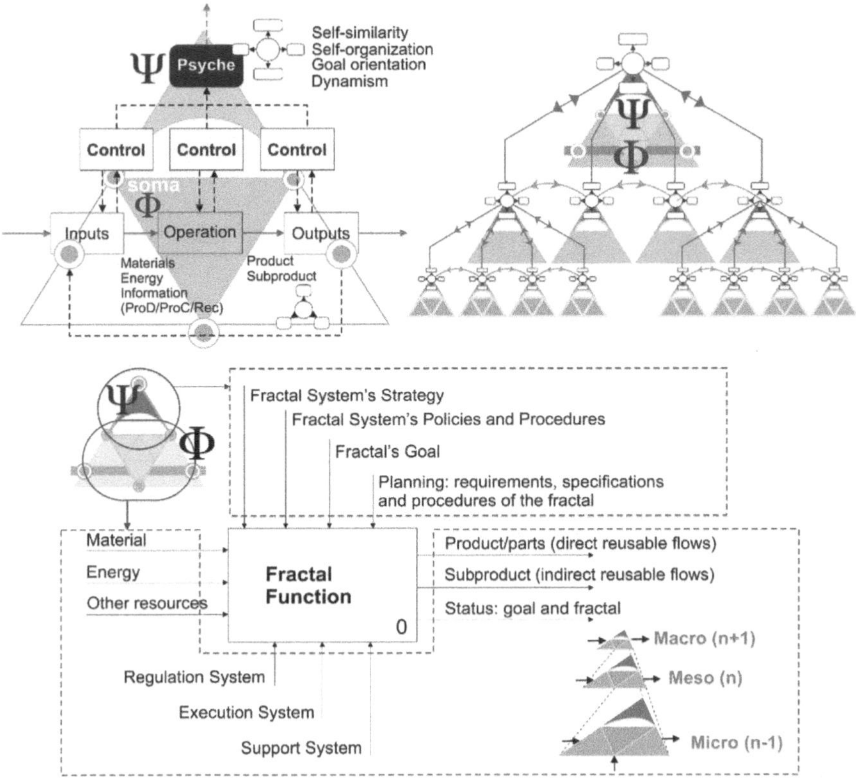

FIGURE 1.4 Structure of control and exchange of data in a sustainable and fractal manufacturing systems.

create CPS (Asare et al., 2016; Jiang et al., 2016; Lee et al., 2015); thus the two characteristic worlds can be identified in the CPS: (1) physical (real results of the operation) and (2) virtual (simulated results of the operation), which will be suitable to control the evolution and self-optimization of the fractal (see Fig. 1.5).

With the resulting functional, operational, and physical requirements, from the psyche of each fractal, the flows of matter and energy that are required to fulfill the execution of the transformation operations are configured. Furthermore, the environmental, economic, and social impact on each operation will be calculated. Fig. 1.6 is a specific example of the data exchange flow and the control carried out by the psyche. Thus each fractal must identify its inputs, outputs, and operations in real time, continuously controlling (1) the direct reusable flows (results: ProD) and those that are not reusable (waste, emissions, fluids: ProD′) that require pretreatment to be converted into subproducts (Mtt). This controls the process of symbiosis of the fractal, in which all the waste

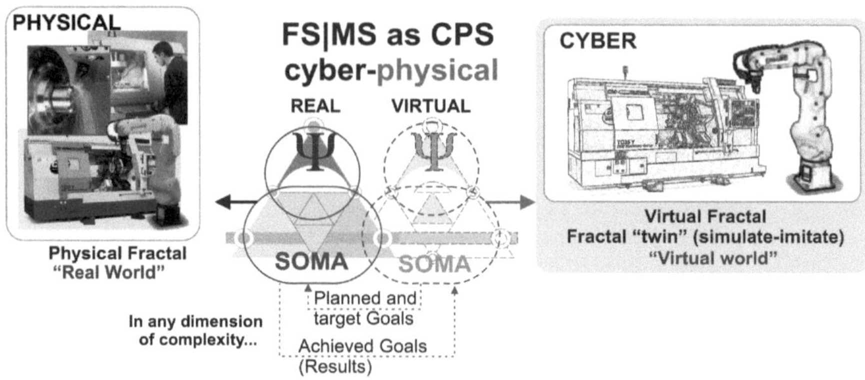

FIGURE 1.5 Fractal cyber-physical system for control of a basic fractal unit.

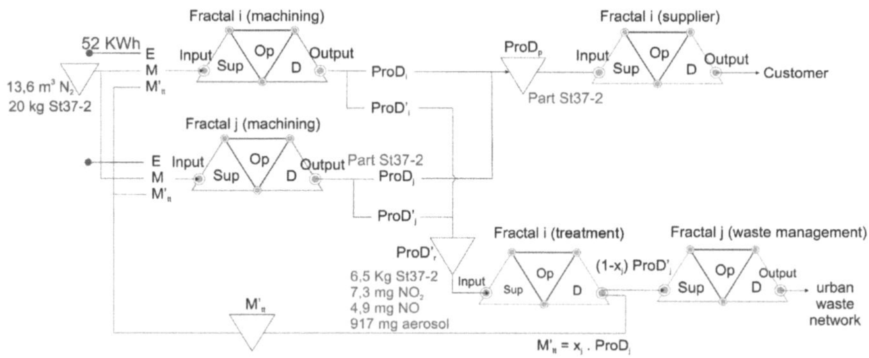

FIGURE 1.6 Example of fractal symbiosis to manufacture a metal part.

must be reduced or converted into reusable inputs [M'tt] for other fractal units with maximum efficiency (minimization of ProD'j).

1.4.5 Complexity dimensions to manage the sustainability results of the sustainable and fractal manufacturing systems

In the process of designing, configuring and implementing a SF|Ms, its structure may be subjected to analysis and synthesis in many aspects (resources, materials, logistics, customer requirements, human resources, manufacturing processes, monitoring and control of the operation, product and subproduct, service, specific inside and outside, layout, etc.). The BFU reduces complexity in the design and management of the fractal (scale invariance) and integrates the analysis, synthesis, and management in a simple manner; but to operate with the self-similarity of the system, it will be necessary to adjust its scale and the level of detail or specificity.

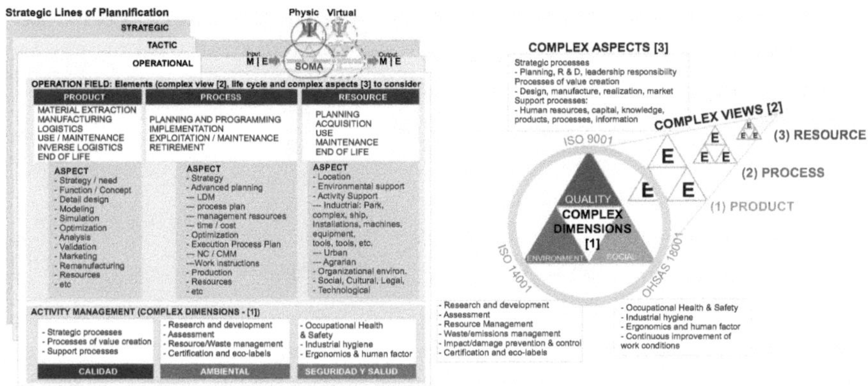

FIGURE 1.7 Views, dimensions, and aspects of complexity in manufacturing.

Thus a tool that manages the successive self-similar structures that form the multiscale and multilevel architecture is required; it proposes a database that will include the complexity dimensions so as to understand all the aspects of complexity that can affect the evolution of the fractal (see Fig. 1.7). Throughout their life cycle they will be able to adapt, modify, or even eliminate themselves, depending on the conditions of the context (self-organization).

1.4.6 Metrics to manage complexity for the sustainable and fractal manufacturing systems

The framework for the SF | Ms includes the assessment of the configuration principles of fractals. Thus three indicators that will compile results in relation to the continuous improvement of the manufacturing system are proposed; they will obtain a weighted result that is linked to finding balanced solutions (economic, social, and environmental), minimizing the complexity of the system.

1.4.6.1 Self-similarity, H(s)

- Metric: $H(s)$
- Principle to apply in the framework: I (principle of scale)
- *Unit*: Static complexity of the fractal's BFU
- *Dimension of evaluation*: Dimensional elements, product, and support/resource
- *Results*: Related to the information and content of the structure (product and support) using the static complexity calculation; it measures the level of the self-similar structure by analyzing information and data of the fractal.

- *Objective*: To quantify the level of the self-similarity between fractal patterns (vertically and horizontally within the structure)
- *Interpretation of results*: The value of the self-similarity is inversely proportional to the complexity of the structure; minimizing the value of $H(s)$ improves the management process of the system.

$$H(s) = -\sum_{i=1}^{M}\sum_{j=1}^{N} p_{ij}log_2(p_{ij}) \tag{1.2}$$

where $H(s)$ = static complexity, S = fractal system (in any level of organization), M = number of data resources (see Figs. 1.2 and 1.3) of the fractal S, N = number of possible different patterns (BFU) in the fractal S, and p_{ij} = probability that a fractal i is in state j (planned pattern)

1.4.6.2 *Coevolution*, H(d)

- *Metric*: $H(d)$
- *Principle to apply in the framework*: II (principle of requisite fractality)
- *Unit*: Dynamic complexity of the fractal
- *Dimension of evaluation*: Mechanisms of action and control (psyche and soma)
- *Results*: Related to the behavior of the fractal (cognition and action) using the calculation of the dynamic complexity
- *Objective*: To quantify the level of deviation from the goals (distance between the planned and the executed) in the behavior of the fractal
- *Interpretation of results*: The value of coevolution is directly proportional to the optimum achievement of objectives; minimizing the value of $H(d)$ improves the processes of self-organization of the fractal, making the system evolve to more optimum states (best adapted to the context) and to improve the results achieved (the objectives are successfully executed according to the plan).

$$H(d) = -PlogP - (1-P)log \quad (1-P) - (1-P)\sum_{i=1}^{M}\sum_{j\in NS_i}^{N} p_{ij}log_2(p_{ij}) \tag{1.3}$$

where $H(d)$ = dynamic complexity, P = probability of the fractal S to be in a planned state (i.e., achieves the planned goals), and p_{ij} = probability that a fractal j is out of control (which depends on the deviation between the actual and planned fractal's goals, i.e., it does not achieve the goal).

1.4.6.3 *Triple-E performance, C_MSTPh*

- *Metric*: C_MS_{TPh}
- *Principle to apply in the framework*: III (principle of ecocompatibility)
- *Unit*: Level of sustainable performance in the economic, environmental and social balance
- *Dimension object of evaluation*: Interaction (impact-damage level) between the SF | FS and its context unit
- *Results*: Efficiency and effectiveness of the metabolic routes (exchange of matter and energy) of the SF | Ms
- *Objective*: To evaluate the management of the metabolic routes with the sustainable performance of the system. Sustainable performance is defined as the level of execution of the fractal in relation to its environmental, social, and economic impact. The performance is measured with the best available techniques that exist, for example, life cycle assessment (Westkämper, Alting, & Arndt., 2001), substance flow analysis (Huang, Vause, Ma, & Yu, 2012), or material flow analysis (Smith & Ball, 2012).
- **Interpretation of results**: To control the metabolic routes, the fractal will regulate its metabolism throughout its life cycle; this process corresponds to the adjustment of its endometabolism and exometabolism (de Molina-Navarro & Toledo, 2014; Garrido, 2007):
  - o *Endometabolism (EndoM)*: A set of input resources (raw materials, energy, water, etc.) that are necessary in order to achieve the objective; it corresponds with the optimum or adjusted level of industrial metabolism.
  - o *Exometabolism (ExoM)*: A set of waste, expenses, and impacts of the fractal that are not necessary to carry out the function but exist as a result of the configuration made by the fractal itself; they contribute to the generation of damage and environmental, economic, and social impact, and they can be minimized with extended lean strategies and other process improvement techniques.

The control of the metabolism of the SF | Ms is carried out by its subsequent reconfiguration (Peralta Álvarez, Marcos Bárcena, & Aguayo González, 2016) incorporating clean technology, best environmental practices, or best available techniques to minimize the endometabolism (selection, consumption, and disposal of inputs, that is, matter and energy, that are strictly necessary for the function) and to reduce, minimize, or eliminate the exometabolism.

A. Quantification of integrated metabolism of the fractal

$$MS = EndoM + ExoM \tag{1.4}$$

where EndoM = endometabolism = biometabolism (environmental impact), ExoM = exometabolism = sociometabolism (social and economic impact).

B. Quantification of the performance level of the adjustment process of the MS_{TPh}:

$$MS_{\text{TPh}} = \frac{EndoM'}{MSx}\alpha = \frac{EndoM'}{EndoM' + ExoM'}x\alpha \tag{1.5}$$

where $MS_{\text{TPh}} = \alpha$, a sustainable goal of the social metabolism, and EndoM = [α, β], a range for which the fractal covers the needs in an ecoefficient and ecoeffective way (optimum and minimum). It is estimated from the virtual fractal of the CPS (Fig. 1.5).

- α is the minimum amount required to satisfy the need.
- β is the maximum amount registered by the fractal in its life cycle to satisfy the need.

EndoM′ = quantification of the endometabolism registered at the moment t of the fractal operation. It is estimated from the real world of the fractal CPS (Fig. 1.5).

ExoM = [0,1], the amount of consumed exometabolism EndoM′ weighted on 1. The objective of the fractal is to achieve ExoM = 0.

- 0 is the nonexistence of exometabolism; MS_{TPh} = EndoM = α.
- 1 is the maximum amount registered by the fractal in its life cycle to satisfy the need; ExoM = EndoM + γ'
- γ is the amount of matter and energy wasted (oversized) in the execution of the activity of the fractal.

EndoM′ = quantification of the exometabolism registered at the moment t of the fractal operation

C.Measurement of complexity of the metabolism:

$$C_{MS_{\text{TPh}}} = \frac{ExoM'}{EndM'} \rightarrow 1 < C_{MS} < \delta \tag{1.6}$$

where δ is the highest quotient registered by the exometabolism/endometabolism ratio in the analyzed period.

The strategies that regulate the exometabolism correspond to the principles of lean manufacturing. They can adjust the endometabolic exchange flows, which are required by the task (materials, energy, and input information, transformed into the expected outputs, products, and subproducts) reducing the seven types of waste that can increase the value of the exometabolism (FEMEVAL, 2011; Matías Hernández & Vizán Idoipe, 2013): transport, inventory, motion, waiting, overproduction, overprocessing, and defects. This strategy aims to expand the scope of lean manufacturing (Peralta Álvarez et al., 2016), in which other types of waste, such as environmental impacts or damage to the health of the worker, are also considered. The extended lean strategies are proposed with the following meaning: The maximization of the value that is created will not only focus on the needs of the client (using minimal resources, adjusted, to obtain quality results) but also be a strategy that is linked to the needs of the natural ecosystem (including social externalities and indirect impacts).

1.5 Conclusion

This chapter presented a section of the proposed bioinspired framework for sustainable fractal manufacturing as a strategy to reduce complexity in the search for balanced sustainable solutions. The chapter began by analyzing the evolution of the complexity management models in manufacturing up to now, identifying the need for a change in the approach that is used when there is an interaction between sustainable and conventional requirements in manufacturing. It proposed the use of fractal theory to reduce sustainable complexity. The framework includes the design and development process of the SF|Ms under the new concept of connected digital manufacturing. The set of synthesis principles, the standard structure of the system (BFU and content structuring of information and data), and an architecture that configures sustainable fractal manufacturing systems as CPSs were defined. In addition, a model that evaluates the complexity and guides the scope of the principles of sustainable functioning in three dimensions (economic, social, and environmental) in the SF|Ms was provided.

References

Aguayo-González, F., Marcos-Bárcena, M., Sánchez-Carrilero, M., & Lama-Ruiz, J. R. (2007). Advanced distributed manufacturing systems. Ra-Ma.

Alfaro, M. D., & Sepulveda, J. M. (2006). Chaotic behavior in manufacturing systems. *Int. J. Prod. Econ.*, *101*(1), 150–158. Available from https://doi.org/10.1016/j.ijpe.2005.05.012.

Anne, K. R., Bhagavatula, S. K., Chedjou, J. C., & Kyamakya, K. (2008). Self-organized supply chain networks: theory in practice and an analog simulation based approach. 30.

Asare, P., Broman, D., Lee, E. A., Torngren, M., & Sunder, S. S. (2016). *Cyber-Physical Systems*. Berkeley CPS Publications.

Ashby, W. R. (1991). *Principles of the self-organizing system. Facets of Systems Science* (pp. 521–536). Springer US. Available from https://doi.org/10.1007/978-1-4899-0718-9.

Attar, A., & Kulkarni, L. (2014). Fractal manufacturing system–intelligent control of manufacturing industry. *Int. J. Eng. Develop. Res.*, 2(2), 1814–1816, https://doi.org/2321-9939.

Baldomero, M., Salmerón, J. L., & Aguayo, F. (2006). Modelado orientado a objetos de un sistema de fabricación fractal. *X Congreso de Ingeniería de OrganizaciónValencia, 13*, 7 y 8 de septiembre de 2006.

Barkman, W. (1989). *In-Process Quality Control for Manufacturing*. CRC Press.

Bauernhansl, T., ten Hompel, M., & Vogel-Heuser, B. (2014). *Industrie 4.0 in Produktion, Automatisierung und Logistik*. Springer. Available from https://doi.org/10.1007/978-3-658-04682-8.

Beaumariage, T., & Kempf, K. (1994). The nature and origin of chaos in manufacturing systems. Proceedings of 1994 IEEE/SEMI Advanced Semiconductor Manufacturing Conference and Workshop (ASMC), 169–174. Available from https://doi.org/10.1109/ASMC.1994.588235

Brandt, D., Hartmann, E., Sander, C., & Strina, G. (1999). Designing and simulating socio-technical systems: concepts and strategies. *Hum. Factors Ergon. Manuf., 9*(3), 245–252, https://doi.org/10.1002/(SICI)1520-6564(199922)9:3 < 245::AID-HFM2 > 3.0.CO;2-X.

Cha, Y., Shin, M., Ryu, K., & Jung, M. (2007). Goal-balancing process for goal formation in the fractal manufacturing system. *Int. J. Prod. Res., 45*(20), 4771–4791. Available from https://doi.org/10.1080/00207540600787234.

Chituc, C., & Restivo, F. (2009). Challenges and trends in distributed manufacturing systems: are wise engineering systems the ultimate answer. Second International Symposium on Engineering Systems, MIT, 1–15.

Chryssolouris, G., Giannelos, N., Papakostas, N., & Mourtzis, D. (2004). Chaos theory in production scheduling. *CIRP Ann. Manuf. Technol., 53*(1), 381–383. Available from https://doi.org/10.1016/S0007-8506(07)60721-5.

de Molina-Navarro, M. G., & Toledo, V. M. (2014). *The Social Metabolism: A Socio-ecological Theory of Historical Change*. Springer International Publishing.

De Toni, A., Nardini, A., Nonino, F., & Zanutto, G. (2001). Complexity measures in manufacturing systems. Proceedings of the European Conference on Complex Systems, 20.

Deshmukh, A. V. (1993). Complexity and chaos in manufacturing systems. In Theses and Dissertations Available from ProQuest (pp. 1–196).

Deshmukh, A. V., Talavage, J. J., & Barash, M. M. (1998). Complexity in manufacturing systems, Part 1: analysis of static complexity. *IIE Trans., 30*(7), 645–655. Available from https://doi.org/10.1023/A:1007542328011.

Dobrescu, R., & Florea, G. (2013). *Unified framework for self-organizing manufacturing systems design, . Service Orientation in Holonic and Multi Agent Manufacturing and Robotics* (472, pp. 57–70). Springer Berlin Heidelberg. Available from https://doi.org/10.1007/978-3-642-35852-4.

Dyke, H. Van, & Irish, B. W. (1985). Fractal actors for distributed manufacturing control. Forthcoming in the Proceedings of the Second IEEE Conference on AI Applications, 653–660.

Efthymiou, K., Pagoropoulos, A., Papakostas, N., Mourtzis, D., & Chryssolouris, G. (2012). Manufacturing systems complexity review: challenges and outlook. *Procedia CIRP, 3*, 644–649. Available from https://doi.org/10.1016/j.procir.2012.07.110.

ElMaraghy, W. H., & Urbanic, R. J. (2004). Assessment of manufacturing operational complexity. *CIRP Ann. Manuf. Technol., 53*(1), 401–406. Available from https://doi.org/10.1016/S0007-8506(07)60726-4.

ElMaraghy, W., ElMaraghy, H., Tomiyama, T., & Monostori, L. (2012). Complexity in engineering design and manufacturing. *CIRP Ann. Manuf. Technol., 61*(2), 793–814. Available from https://doi.org/10.1016/j.cirp.2012.05.001.

FEMEVAL. (2011). Control de Gestión. *Lean Process y Reingeniería*, 59.

Frizelle, G., & Woodcock, E. (1995). Measuring complexity as an aid to developing operational strategy. *Int. J. Oper. Prod. Manag., 15*(5), 26–39. Available from https://doi.org/10.1108/01443579510083640.

Fryer, P., & Ruis, J. (2004). What are fractal systems? (Complex Adaptive and Emergent Systems). Fractal.Org.

Gálvez-Medina, E., Balankin, A., & Resenos, E. (2009). Enfoque fractal aplicado en la administración. *Investig. Admin.*, *43*(1).

Garrido, F. (2007). El paradigma ecológico en las ciencias sociales. Icaria Editorial.

Horns, A. (1989). Job shop control under influence of chaos phenomena. Proceedings. IEEE International Symposium on Intelligent Control, 227–232. https://doi.org/10.1109/ISIC.1989.238689

Huaccho Huatuco, L., Efstathiou, J., Sivadasan, S., & Calinescu, A. (2001). The value of dynamic complexity in manufacturing systems. Proceedings of the International Conference of the Production and Operations Management Society, 180–188.

Huang, C.-L., Vause, J., Ma, H.-W., & Yu, C.-P. (2012). Using material/substance flow analysis to support sustainable development assessment: a literature review and outlook. *Resour. Conserv. Recycl.*, *68*, 104–116. Available from https://doi.org/10.1016/j.resconrec.2012.08.012.

IDE (2016). Industria 4.0. Claves para competir en el futuro.

Jiang, P., Ding, K., & Leng, J. (2016). Towards a cyber-physical-social-connected and service-oriented manufacturing paradigm: social manufacturing. *Manuf. Lett.*, *7*, 15–21. Available from https://doi.org/10.1016/j.mfglet.2015.12.002.

Karaçali, Ö., Halil, İ., & Demirci. (2009). Technology application of chaos theory to data modelling method based on the alternative scenarios in manufacturing information systems. *Technology (Elmsford, N.Y.)*, *12*(2), 107–115.

Kühnle, H. (2009). *Self-similarity and criticality in dispersed manufacturing: a contribution to production networks control. Dispersed Manufacturing Networks. Challenges for Research and Practice* (pp. 59–76). London: Springer. Available from https://doi.org/10.1007/978-1-84882-468-3_4.

Kumar, V. (1987). Entropic measures of manufacturing flexibility. *Int. J. Prod. Res.*, *25*(7), 957–966. Available from https://doi.org/10.1080/00207548708919888.

Kwangyeol, R. (2003). Fractal-based reference model for self-reconfigurable manufacturing systems.

Lee, J., Bagheri, B., & Kao, H.-A. (2015). A cyber-physical systems architecture for Industry 4.0-based manufacturing systems. *Manuf. Lett.*, *3*, 18–23. Available from https://doi.org/10.1016/j.mfglet.2014.12.001.

Markfort, D., Martinetz, J., Klostermeyer, A., & Brehmer, N. (2002). Advanced fractal companies use information supply chains (ADRENALIN). October 2000, 0–40.

Matías Hernández, J. C., & Vizán Idoipe, A. (2013). Lean manufacturing Conceptos, técnicas e implantación. Fundación eoi.

Mattsson1, S., Gullander2, P., Davidsson, A., n.d. Method for measuring production complexity.

McCarthy, I. P., Rakotobe-Joel, T., & Frizelle, G. (2000). Complex systems theory: implications and promises for manufacturing organisations. *Int. J. Manuf. Technol. Manag.*, *2* (1–7), 559–579.

Moody, P. E., & Morley, R. E. (1999). *The Technology Machine: How Manufacturing Will Work in the Year 2020*. Free Press.

Morgan, G. (1996). Imágenes de la organización. Alfaomega.

Morley, R. (1996). Painting trucks at general motors: the effectiveness of a complexity-based approach. Embracing Complexity: Exploring the Application of Complex Adaptive Systems to Business, 53–58.

Papakostas, N., Efthymiou, K., Mourtzis, D., & Chryssolouris, G. (2009). Modelling the complexity of manufacturing systems using nonlinear dynamics approaches. *CIRP Ann. Manuf. Technol.*, *58*(1), 437–440. Available from https://doi.org/10.1016/j.cirp.2009.03.032.

Peralta Álvarez, M. E., Marcos Bárcena, M., & Aguayo González, F. (2016). A review of sustainable machining engineering: optimization process through triple bottom line. *J. Manuf. Sci. Eng.*, *138*(10), 100801. Available from https://doi.org/10.1115/1.4034277.

Peralta, M. E., Marcos, M., & Aguayo, F. (2016). On the sustainability of machining processes. Proposal for a unified framework through the triple bottom-line from an understanding review. *J. Clean. Prod.*, *142*(Part 4), 3890–3904. Available from https://doi.org/10.1016/j.jclepro.2016.10.071.

Peralta, M. E., Marcos, M., Aguayo, F., & Lama, J. R. (2017). Advanced fractal manufacturing: multi-level and multi-scale proposal for sustainable manufacturing processes. *Int. J. Mechatron. Manuf. Syst.*, *10*(1), 3. Available from https://doi.org/10.1504/IJMMS.2017.084375.

Peralta, M. E., Marcos, M., Aguayo, F., Lama, J. R., & Córdoba, A. (2015). Sustainable fractal manufacturing: a new approach to sustainability in machining processes. *Procedia Eng.*, *132*, 926–933. Available from https://doi.org/10.1016/j.proeng.2015.12.579.

Peralta-Alvarez, M. E. (2016). Modelado, simulación y optimización de la sostenibilidad de procesos de fabricación para su integración multiescala y multinivel. Aplicación a células de mecanizado. Universidad de Cádiz.

Poincare, H. (1946). La ciencia y método.

Ryu, K., & Jung, M. (2003). Dynamic restructuring process in the fractal manufacturing system. Proceedings of the 2003 KIIE Fall Conference, 54–58.

Ryu, K., & Jung, M. (2004). Goal-orientation mechanism in a fractal manufacturing system. *Int. J. Prod. Res.*, *42*(11), 2207–2225. Available from https://doi.org/10.1080/00207540410001661427.

Ryu, K., Son, Y., & Jung, M. (2003). Modeling and specifications of dynamic agents in fractal manufacturing systems. *Comput. Ind.*, *52*(2), 161–182. Available from https://doi.org/10.1016/S0166-3615(03)00099-X.

Sauter, R., Bode, M., & Kittelberger, D. (2015). How Industry 4.0 is changing how we manage value creation.

Shin, M., Mun, J., & Jung, M. (2009). Self-evolution framework of manufacturing systems based on fractal organization. *Comput. Ind. Eng.*, *56*(3), 1029–1039. Available from https://doi.org/10.1016/j.cie.2008.09.014.

Shin, M., Ryu, K., & Jung, M. (2012). Reinforcement learning approach to goal-regulation in a self-evolutionary manufacturing system. *Expert Syst.Appl.*, *39*(10), 8736–8743. Available from https://doi.org/10.1016/j.eswa.2012.01.207.

Silva, N., & Ramos, C. (1999). Proposal for inter-enterprises negotiation infra-structures using an holonic approach; Nuno Silva and Carlos Ramos. Proceedings of 1st International IFAC Workshop on Multi-Agent Systems in Production, 2–4.

Simon, H. (2006). Las ciencias de lo artificial. Comares.

Smith, L., & Ball, P. (2012). Steps towards sustainable manufacturing through modelling material, energy and waste flows. *Int. J. Prod. Econ.*, *140*(1), 227–238. Available from https://doi.org/10.1016/j.ijpe.2012.01.036.

Strauss, R. E., & Hummel, T. (1995). The new industrial engineering revisited-information technology, business process re-engineering, and lean management in the self-organizing "fractal company." Proceedings for Operating Research and the Management Sciences, 287–292. Available from https://doi.org/10.1109/IEMC.1995.524595

Tharumarajah, A. (1996). Comparison of the bionic, fractal and holonic manufacturing system concepts. *Int. J. Comput. Integrate. Manuf.*, *9*(3), 217–226. Available from https://doi.org/10.1080/095119296131670.

Tharumarajah, A., Wells, A. J., & Nemes, L. (1998). Comparison of emerging manufacturing concepts. SMC'98 Conference Proceedings. 1998 IEEE International Conference on Systems, Man, and Cybernetics (Cat. No.98CH36218), 1, 325–331. Available from https://doi.org/10.1109/ICSMC.1998.725430

Tirpak, T. M. T. M., Daniel, S. M. S. M., LaLonde, J. D. J. D., & Davis, W. J. W. J. (1992). A note on a fractal architecture for modelling and controlling flexible manufacturing systems. *IEEE Trans. Syst. Man Cybern.*, *22*(3), 564–567. Available from https://doi.org/10.1109/21.155958.

Vega, V. B. E. (2012). Structural complexity of manufacturing systems layout. 120.

Warnecke, H.-J. (1993). *The Fractal Company: A Revolution in Corporate Culture*. Springer-Verlag Berlin Heidelberg. Available from https://doi.org/10.1007/978-3-642-78124-7.

Weichhart, G. (2013). The learning environment as a chaotic and complex adaptive system. E-Learning Support for Thrivability. 1(1).

Westkämper, E., Alting., & Arndt. (2001). Life cycle management and assessment: aproaches and visions towards sustainable manufacturing. *Proc. Inst. Mech Eng. B J Eng Manuf.*, *215*(5), 599–626.

Wiendahl, H.-P., & Harms, T. (2004). Self-organized production based on agent theory. *Prod. Plan. Control*, *15*(2), 213–224. Available from https://doi.org/10.1080/09537280410001662493.

Xu, J., & Chen, C. (2010). Similarity in tacit knowledge of supply chain and its inferences. 2010 International Conference on E-Product E-Service and E-Entertainment, 1–5. Available from https://doi.org/10.1109/ICEEE.2010.5660172

CHAPTER

2

Reliability of manufacturing technologies

Marius Gabriel Petrescu, Costin Ilincă and Maria Tănase

Petroleum-Gas University of Ploieşti, Ploieşti, Romania

2.1 Introduction

To have the initiative to manufacture a product or offer a service not only requires a solid knowledge of manufacturing technology, but also requires knowledge in many more fields, such as engineering, economics, and environmental protection. At this stage of human development it can be said that technology loses its primary significance—that of ensuring the transformation of raw materials into products—and acquires much more subtle values that are related to ensuring the safety of life for both the direct consumer and all the people and/or nature encountered along the route from resources to product. The designer and the technological engineer are not only concerned with ensuring the correctness and continuity of the manufacturing processes. They must find alternatives to the material resources that are used in manufacturing. They must ensure the possibility of reusing the waste that is produced during manufacture. They must give the product, beyond the characteristics of use, the ability to be "friendly" to the consumer and the environment. They must assume responsibility for the use of the product for reintroduction into the industrial circuit. This circuit does not exclude distributors and consumers, who, in turn, have the same duties as manufacturers. We deduce from this that we are all parties in an industrial/economic circuit, and the proper functioning of the system depends on each of us.

Innovation and Sustainable Manufacturing
DOI: https://doi.org/10.1016/B978-0-12-819513-0.00007-X

2.2 Technologies in the context of the sustainable economy

The concept of technology is traditionally used in the context of product manufacturing processes and the provision of a service. Thus technology can be defined as follows:

- The science of the processes and means of material processing; the set of processes, procedures, and operations that are used to obtain a certain product.
- The totality of the machinery, equipment, procedures, and methods that are used to manufacture the assembly that allows the extension of the action capacity of the operator.
- The science of manufacturing processes and equipment that allow the transformation, in technical and economic conditions, of raw materials into products.
- The ability to use, in a local context and for a specific purpose, the sciences, methods, and equipment to design and manufacture a product and its manufacturing process.

As can be seen, manufacturing technology is inextricably linked to the processes of transformation of raw materials or materials. If we extend the meaning of the notion of transformation, we can see that all the processes that manifest in the life cycle of products, regarding any transformation that occurs from the design phase to consumption, are characterized by technologies (technological processes) with good roles and purposes defined. Thus we can talk about design technologies, manufacturing technologies, distribution technologies, consumption technologies, collection technologies, recovery technologies, and so on. Consumption technology itself can be extended to all other processes if we take into account the fact that in the economic circuit, the processes are performed within systems that alternately play the roles of supplier and beneficiary (customer or consumer) (Fig. 2.1).

The consumer is the person or other entity that purchases and consumes material goods that result from a previous process to meet needs such as production needs, esthetic needs, and physiological needs.

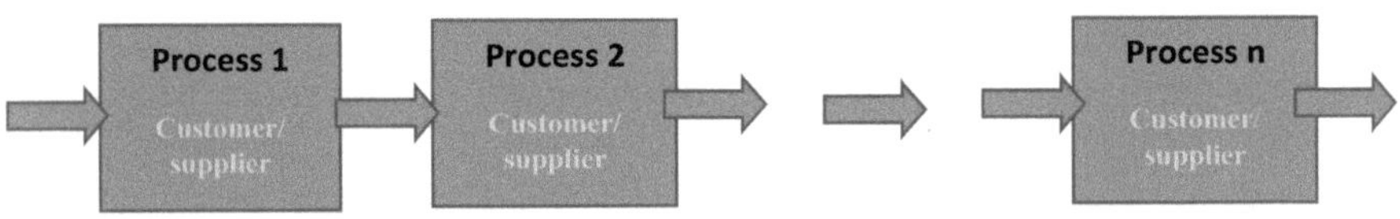

FIGURE 2.1 The dual role of processes in the customer/supplier economy.

2.2.1 Consumption technology

In general, a product has a well-defined destination, which is intended to meet predetermined needs. However, the mode and degree of products use depends on each consumer. Consequently, the transformations, in detail, supported by the products depend on the each consumer profile. The main result of consumption is the consumer satisfaction. The reality shows that any consumption of products (consumption does not refer exclusively to the use of finished products or applications but also includes the consumption of raw materials, materials and resources that represent inputs in any process) generates unusable results at the place of consumption, known by the term "waste." If we take into account all the aspects previously presented, then in economic practice, we can talk about a consumption technology.

In modern society, which, in addition to being a consumer society, must be an efficiency society, the results of the consumption of goods and services must be incorporated into waste recovery technologies. In a complete technology, which foresees, beginning in the design phase, the role and importance of recovering the results after consumption, it is necessary to specify the role of each process corresponding to its place in the circuit of transformations and highlight the individual contribution to a sustainable construction. In this context, the role of the sales and consumption stages must also be noted in order to be included in the loop of the economic system. A sustainable manufacturing technology, to contribute to the generation of a sustainable economy, must prove a complete cycle (Fig. 2.2). Such a technology must prove its reliability. The overall reliability of the technology is dependent on the reliability of each component, starting with the design and ending with the processing of the consumption results, How we can promote the reliability of technologies, what the possibilities are to support them, and what current concerns are related to this topic, we will explore in this chapter.

To refer to consumption in the context of this chapter involves taking into account all the phases and technologies which are found in the manufacturing-use cycle of a product, potentially generating waste (Fig. 2.2).

If we refer strictly to the consumer, practice shows that the efficiency of consumer technology reflects the extent to which these rules are followed:

- The person must control the product (implicitly the consumption technology of the product) and not vice versa.
- The product, no matter how advanced or efficient it is, cannot 100% satisfy the consumer's needs.
- The impact that the product and the consumer technology have on the consumer depends exclusively on the human being.

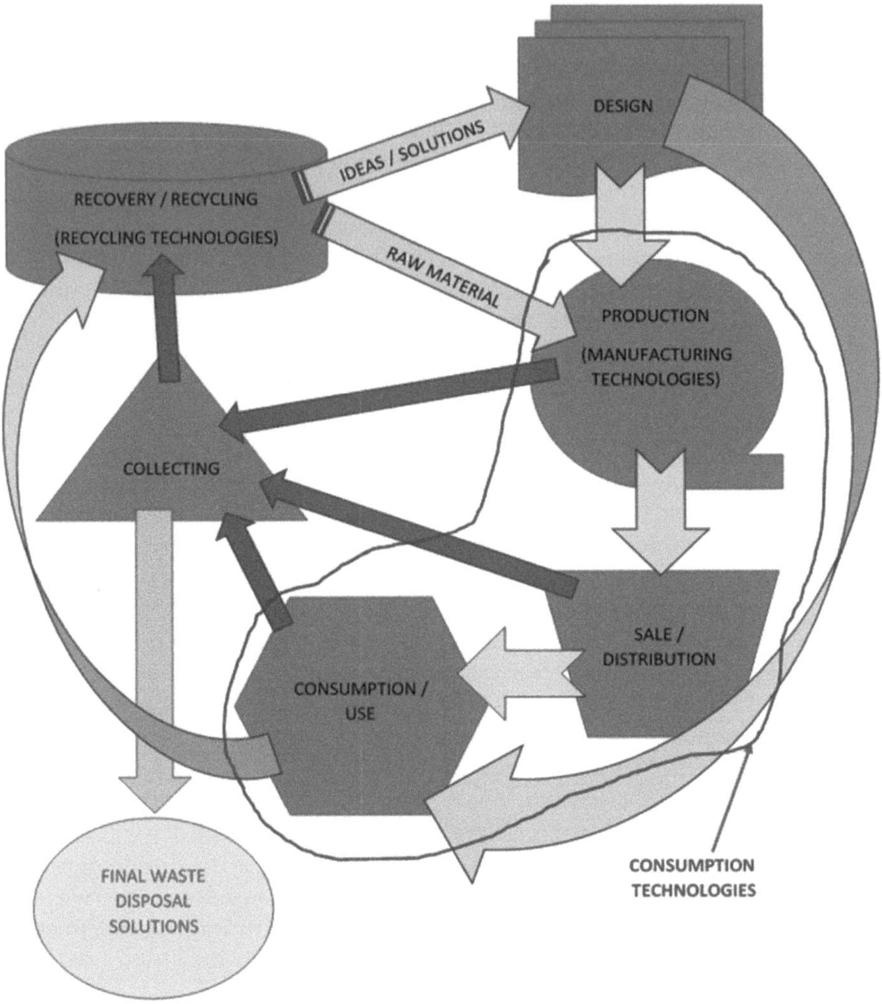

FIGURE 2.2 The place and the role of the participants (phases, stages, technologies) in promoting a complete technology in support of the idea of a circular economy.

What can be deduced from this?

First, we must avoid consumption of products that are not absolutely necessary. At the moment, there is a tendency on the part of consumers to buy products that satisfy a trend rather than real consumption needs. This generates residual components (waste) in excessively large quantities.

Second, we need to apply selective and differentiated consumer technologies, in the sense of identifying the essential requirements of consumers. Differentiated consumption technologies, in the sense of assimilating specialized products, have a high degree of satisfaction of individual requirements to the detriment of multifunctional products, which offer a compromise between the levels of satisfaction that are characteristic of various functions and attributes.

Third, it is possible to discuss education of the consumer in the sense that the consumer must acquire the skills to obtain from the product, within the consumer technology, the maximum level of satisfaction or use. The immediate consequence of this is a minimum level of waste.

To meet these requirements, it is necessary to integrate, within the circular economy (see Fig. 2.2), innovative manufacturing strategies (companies and, in general, the economic environment) and optimized consumption technologies, both coupled with efficient recycling technologies. Compliance with these conditions ensures the worth of each technology and the sustainability of the entire cycle.

2.3 Industrial innovation and manufacturing innovation

The market situation is the one that forces industrial producers to generate strategies that are capable of ensuring their companies' sustainability. Managers must analyze the position of their companies in the economic landscape and make decisions in correlation with the trends of the field in which they operate. The modern economy relates to five fundamental landmarks:

1. Resource availability.
2. The categories of products and services requested by the company.
3. The efficiency of production technologies.
4. The categories of customers and their expectations.
5. Balance between the market requirements, available resources, and environmental protection.

To face these challenges, the company management must show flexibility when formulating its policies, and the strategies must refer to actions such as the following:

- Performing analyzes that design different scenarios based on both current and prospective economic policies, including constraints (internal to the company but also external, specific to the environment in which the company operates).
- Planning medium- and long-term activities and prioritizing actions that give flexibility to production.
- Elaboration of mechanisms to ensure agility and flexibility in the relationship with business partners (suppliers, collaborators, customers).

- Managing the multifunctional impact of changes of any nature in the environment (economic, political, social, etc.).
- Rigorous inventory of all categories of resources, their value, and the possibilities of recovery or reuse in the consumption cycle.
- Promoting communication with customers to identify the desire to move to the production of new products, to sell products at a certain price, or even to redesign the business model.

As can be seen in going through these proposals, maintaining a company in the market requires a special management dynamism. The set of changes and transformations that are specific to a modern development strategy are generated under the sign of innovation: product innovation, process innovation, and innovation of company policies. Innovation is essential for a company's growth strategy. Moreover, innovation is essential for the survival and growth of technologies.

Economic studies show that the idea that dynamic industries are constantly growing on a large scale may be—or really is—entirely wrong (https://hbr.org/2015/06/you-need-an-innovation-strategy). Many of the industries that include advanced technologies, so-called growth industries, include segments that do not grow at all. At the same time, it is found that industries that are considered mature, such as consumer goods, often have rapidly growing segments. Similarly, even industrial sectors that are perceived as "emerging" usually have, in addition to maturing sectors, other sectors that are still in the market expansion phase (https://hbr.org/2015/06/you-need-an-innovation-strategy). The explanation can be found in the fact that economies show an extreme dynamism—not necessarily mandatory by growth but especially by fluctuations and changes in objectives—and, equally, customers frequently change their options.

To ensure company sustainability (products or manufacturing), it is necessary to encourage innovation. This requires a culture of innovation at the organizational level. The role of an organization's management is to facilitate innovation within the company.

A culture of innovation refers to the company's ability to make use of all the resources that can lead to innovation, especially the employees. To encourage ideation and creation, companies need to rebuild their hierarchical systems so that innovation thrives (https://innovatingsociety.com/inovatia-in-pasi-simpli/).

Innovation is specific to the organization as a whole or can manifest only in the process or product. The choice is at the discretion of management and depends on the objectives that have been set at the organization level.

Product innovation is based mainly on the study of customer requirements and expectations (which are undeclared requirements). The result

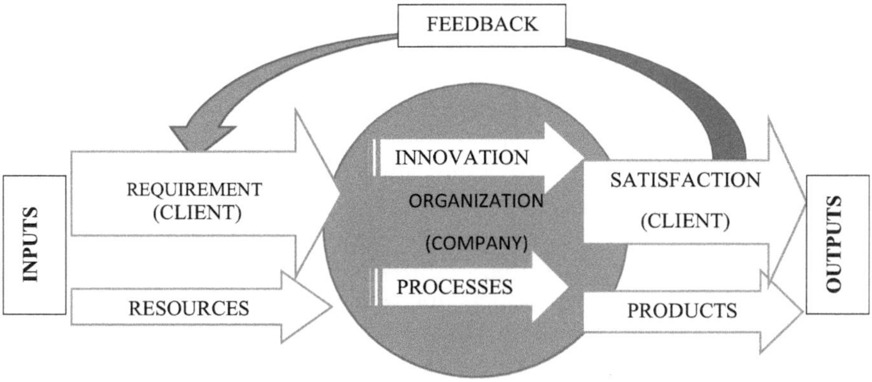

FIGURE 2.3 The place and role of the consumer (customer) in defining the company's strategy (authors' proposal).

of product innovation is an improved product or a new product that is radically different from the previous one.

Process innovation is the improvement of the process and is based mainly on the adoption of new manufacturing technologies or adaptation the existing technologies to the constraints imposed by the environment: reducing the consumption of raw materials, stocks reduction, recycling and reuse of waste, pollution reduction, and so on. The result of process innovation is not necessarily a new product or service but is a contribution to manufacturing sustainability.

Innovation at the organization level can include product and process innovation. It can also refer to changing the managerial or business methods.

An innovation strategy must not be subordinated to the idea of market capture. In establishing the innovation strategy, the relationship between the entrances and exits of the organization must be correctly understood (Fig. 2.3). The place and role of the consumer (customer) in defining the company's strategy in general and the innovation strategy in particular must be understood.

2.4 The challenges of sustainable economic systems

Implementation of the concept of a sustainable economy has given rise to new challenges regarding the identification of solutions to reduce the pressure exerted by production and consumption on resources and the environment. The continuous increase in demand for products and services has always stimulated production. The new conditions have forced the transition to sustainable production and, equally, the development of sustainable growth, whether it is an increase in consumption or profit.

The classic model of the economy was that of procurement-production-disposal, a linear model (Fig. 2.4), through which each product has a limited life span. Such a model mainly favored the manufacturers, who promoted frequent modernization and state-of-the-art technologies that were thought to be absolutely necessary and developed products with a limited, often short life span so as to encourage replacement consumption. The consumed products, according to this model, are to be disposed of as waste. Regardless of the economic category—food, civil engineering, household appliances, automobiles, energy production—the products include raw materials and valuable materials, which are found in nature in limited quantities. Their disposal, as a result of a productive action or as a result of consumption, in the form of waste not only represents a loss of raw material, but also leads to environmental degradation. As a result of such a practice, we are witnessing increases in the prices of metals and minerals, fossil fuels, feed and food, drinking water, fertile land, and so on.

Every year, the European Union uses almost 15 tons of materials per person, with every EU citizen generating an average of more than 4.5 tons of waste a year, almost half of which is dumped in landfills. The linear economy, based exclusively on resource extraction, is no longer a viable option (http://epp.eurostat.ec.europa.eu/portal/page/portal/environment/introduction).

The current context calls for solutions that extend the products' life and lead to the creation of competitive products that last over time (in working order or in the form of reusable components). Thus cars, computers, household appliances, packaging, and other products may be redesigned, taking into account the requirements of strength, reuse, repair, remanufacturing, and recycling (http://publications.europa.eu/resource/cellar/c8cfd1ae-6285-40ba-879ff2e78e4c2b6e.0013.02/DOC_1).

The concept of a sustainable economy cannot be limited to products. It should not be overlooked that products are the result of design and

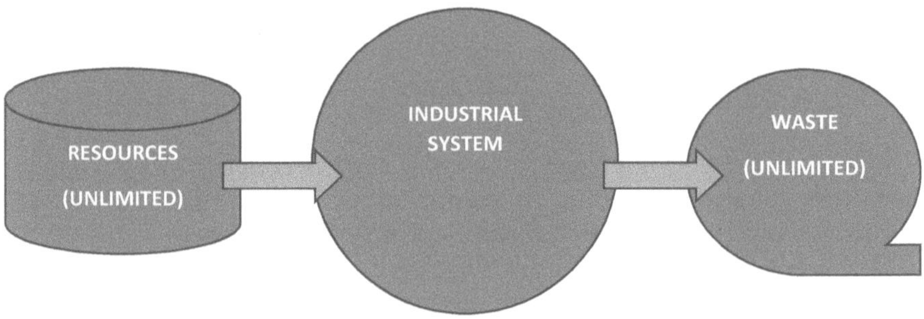

FIGURE 2.4 The linear model of industrial systems.

manufacturing processes in which the products' use properties are defined. In addition to the fact that the development of recyclable products begins in the design and manufacturing stages, the manufacturing technologies themselves must be sustainable.

Promoting sustainable manufacturing can be achieved through improved manufacturing technologies that include alternative green manufacturing processes. Green manufacturing technologies have been found to have a high potential to reduce costs and improve competitiveness by reducing resource consumption. Therefore by generating less waste and consuming less energy, they will have a lower social and environmental impact (http://iosud.utcluj.ro/files/Dosare%20abilitare/Fratila%20Domnita/TEZA%20ABLT_Fratila%20D_martie%202019.pdf).

The sustainable alternative to conventional manufacturing technologies is represented by environmentally friendly manufacturing technologies (Fig. 2.2), which provide an improvement in the process performance, the system—whether we are referring to the production/manufacturing system or the economic system as a whole—and the final product by applying the 6R concept: reduction, reuse, recovery, recycling, reconditioning, and redesign (http://iosud.utcluj.ro/files/Dosare%20abilitare/Fratila%20Domnita/TEZA%20ABLT_Fratila%20D_martie%202019.pdf).

It is obvious that ensuring the sustainability (durability) of manufacturing technologies is a difficult problem to solve in the conditions of multiple constraints that the system manifests (Fig. 2.5). In fact, this is an optimization problem that involves identifying solutions that are close to the maximum and minimum levels of the parameters from case to case (Table 2.1).

The directions and the action ways set out earlier in relation to manufacturing products and technologies are, in fact, the basis for the concept of a circular economy. In a circular economy, any new product or service is designed according to the principle of circularity. Increasing the level of cooperation in and between the manufacturing processes and supply chains, as well as between traders and customers, can reduce costs, waste, and harmful effects on the environment.

Currently, manufacturing systems are undergoing a transition from the linear operating model (Fig. 2.1) to the semicircular model. This model is distinguished by much more efficient waste management, which results in a reduction in the amount of waste generated by the system. In a circular economy the model multiplies in each of the component systems (Fig. 2.6).

The reasons for the emergence and promotion of the concept of a circular economy are mainly the following:

- Resource efficiency. This does not refer exclusively to care for the environment but takes into account all aspects related to the

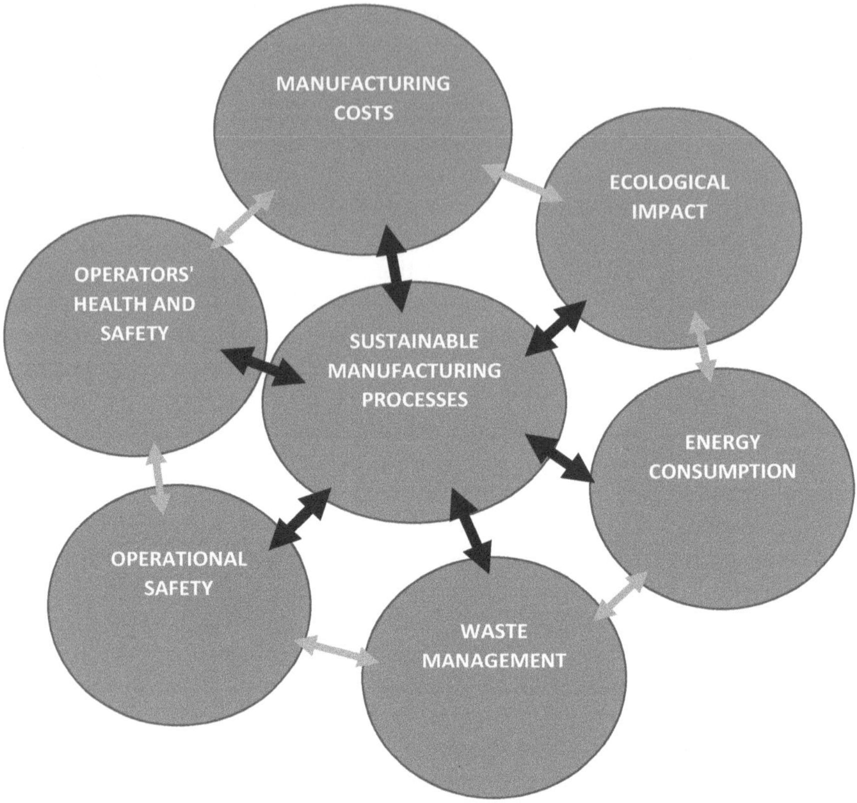

FIGURE 2.5 The conditioning of sustainable manufacturing processes (http://iosud.utcluj.ro/files/Dosare%20abilitare/Fratila%20Domnita/TEZA%20ABLT_Fratila%20D_martie%202019.pdf).

TABLE 2.1 The objectives of designing sustainable technologies.

Constraint	Target level
Energy consumption	Minimum
Manufacturing costs	Minimum
Ecological impact	Minimum
Operational safety	Maximum
Operators' health and safety	Maximum
Waste management (volume of waste)	Minimum

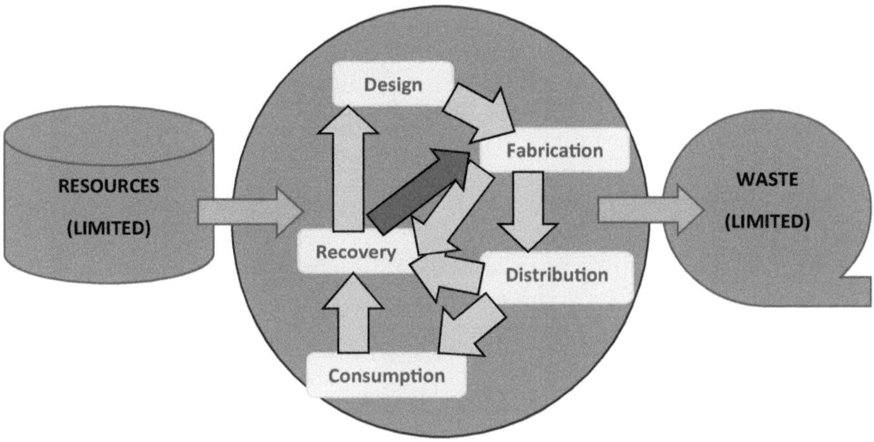

FIGURE 2.6 The semicircular model of industrial systems (semicircular economy).

efficiency of the processes that a product transits. The efficiency criterion is, without a doubt, decisive in assessing the system's performance (in particular, organizations and companies) and processes (including the technological ones).

- Increasing competitiveness. There is a causal link between competitiveness and the economic growth of a system or process in the sense that there can be no sustainable economic growth without proper competitiveness (http://dspace.usarb.md:8080/jspui/bitstream/123456789/2335/1/Bogdan_R_competitivitate_economica.pdf).
- Economic security and energy independence. Economic security refers to ensuring the organization's ability to adapt to the external environment signals (market), that is, to opportunities but also to the risks associated with its manifestation in conditions of performance and competitiveness. Energy independence means ensuring the necessary energy from the organization's own sources, including waste.
- Equal opportunities and equity. These imply the imposition of the same rules for all participants in the circular economy.

Technological optimization will lead to a reduction in the consumption of energy and natural resources while reducing the amount of waste. As the concept of a circular economy is implemented, the functioning of industrial systems will increasingly disrupt their external systems. Ideally, the outputs from one system will constitute inputs into another system, and consequently, the notions of resources and waste will lose their meaning (Fig. 2.7).

The transition to a circular economy focuses on the reuse, repair, reconditioning, and recycling of materials and used products. The circularity transforms what was considered waste into a resource.

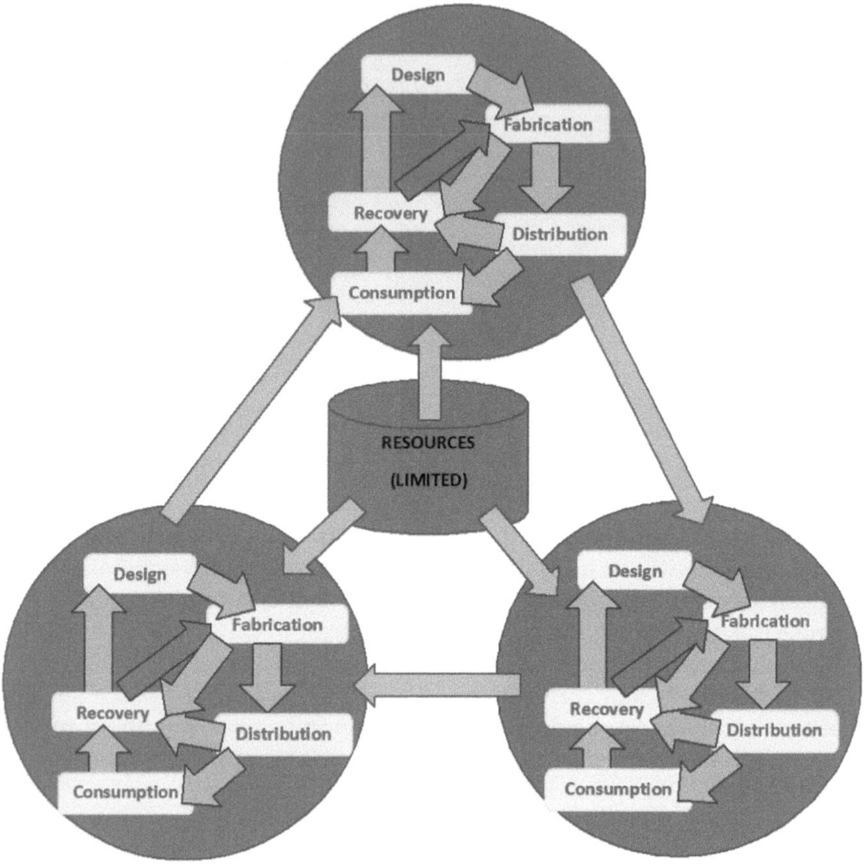

FIGURE 2.7 The circular model of industrial systems (circular economy).

This approach should not scare manufacturers, as it gives a new meaning to innovation in the form of "eco-innovation." They will discover new opportunities in the development of new products, processes, technologies (repair and/or recycling), and organizational structures.

As can be seen, a circular economy involves, on the one hand, stimulating creative activity (eco-innovation in design and production) and, on the other hand, encouraging consumer participation in closing the product and waste circuit.

As is shown in the schematic representation in Fig. 2.8, a circular economy system is based on the integration of waste in the classical circuit of production: raw materials → processing → consumption, not in the form of an appendix but in the form of a link that closes the loop (in reality, several loops are formed to reintegrate the waste into production; see Figs. 2.5 and 2.7). Consequently, the cycle is repetitive and looks like this: raw materials

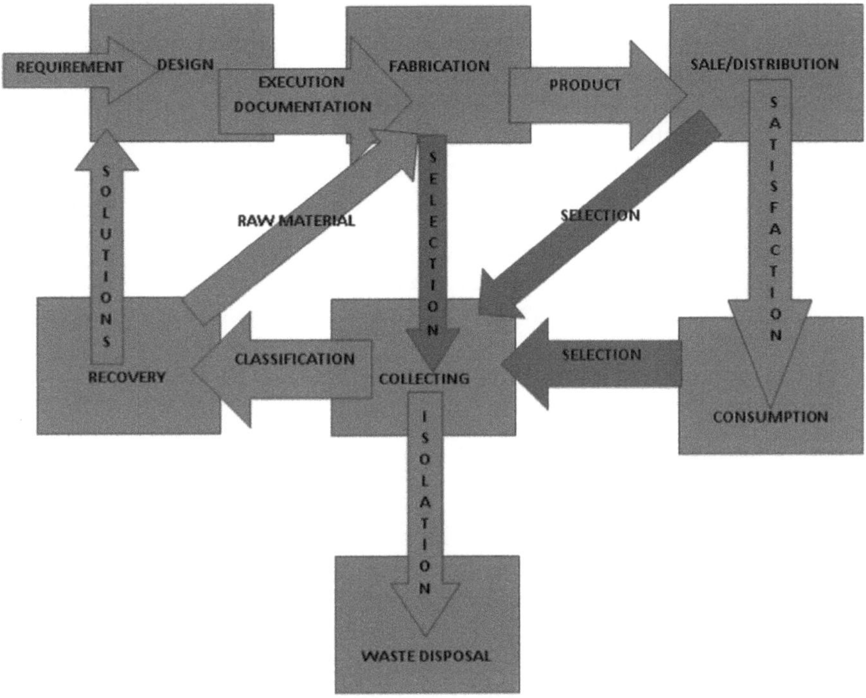

FIGURE 2.8 The systemic (process-based) representation of a circular economy.

→ processing → consumption → reuse of waste in production (https://vtex.com/ro/blog/trenduri/modelul-de-economie-circulara-cele-mai-bune-practici-pentru-ecommerce/ or https://vtex.com/en/blog/trends/running-a-sustainable-ecommerce-a-higher-priority-than-ever/ accessed 22.10.2021).

Such an approach requires organizations to commit to focusing as much on the social and environmental concerns as they do on profit. The primary goal of companies—the profit—must coexist with specific goals of the sustainable economy, which focuses on people and the environment. Best of all, this reorientation is described by the triple bottom line (TBL) theory. TBL recommends to the companies to work simultaneously on these three directions/baselines: Profit—traditionally expressed through profit and loss; People—the measure of the organization social responsibility throughout its existence; Planet—the measure of the company's responsibility to the environment (https://www.investopedia.com/terms/t/triple-bottom-line.asp). TBL theory draws attention to the need for social involvement of organizations (corporate social responsibility) in order to capture the overall picture of the economic environment with real involvement on the total business cost (https://www.investopedia.com/terms/t/triple-bottom-line.asp).

All these comments have the role of providing a complete picture regarding the complexity of the issues to be considered when we aim to analyze the sustainability of the circular economy in terms of reliability and risks associated with the system and processes.

Adhering to the circular economy idea means being aware of the opportunities but also the risks that such an option entails.

In the category of opportunities it is worth mentioning the following:

- Increasing the competitiveness.
- Cost reduction in the context of streamlining processes.
- Strengthening the partnerships that are established between suppliers and beneficiaries.
- Customer loyalty by adhering to the organization's policies.
- Improving the organization's image.

The specific risks of integration into the circular economy relate mainly to the following:

- Risks of legal noncompliance, which can be translated into additional costs for compliance with legal provisions.
- Isolation from customers and the community, given that the external environment adheres to the circular economy concept.
- Increasing environment-related taxes in order to discourage deviations from the principles of the circular economy.
- Escalation of the risk level by cascading the risks.

2.5 The reliability of technologies (processes) in the context of the circular economy

The reliability analysis of the circular economy system must start from the identification of nonconformities that may manifest during the implementation of the processes that contribute to the circularity of resource use. These nonconformities will generate undesirable effects that will negatively affect the idea of circular economy. Each process in the economic circuit will influence the reliability of the circular economy system and, to the same extent, will be influenced by the other processes (technologies) in terms of the integration efficiency in the circuit. To increase the reliability of the whole system, action must be taken on the causes that may lead to possible nonconformities in order to reduce the effects produced by them. It follows that the reliability analysis of the circular economy system, which in a broad sense can be assimilated into a complex technological process, must be based on the results of a risk analysis applied to the processes, subsystems, and technologies that are involved: design, production, distribution, consumption, collection, and recovery and recycling.

Table 2.2 presents a proposal for the identification of possible nonconformities that may characterize the circular assembly of these processes. It must be said that the issues mentioned in Table 2.2 are not the only ones that can be analyzed, just as they are not mandatory for all the industrial systems that are encountered in economic reality. For real industrial systems the analysis may be limited to certain types of nonconformities or may include new aspects that are specific to the analyzed system. The role of the example is to draw attention to the fact that the processes in the economic circuit are naturally interdependent, not only because of the supplier-beneficiary relationship but also as a result of identifying the common causes in the manifestation of nonconformities.

For particular industrial systems the risk analyzes will be based, by obligation, on real data that are collected during their operation or similar data (in the case of predictive risk analyzes, performed in the design phase of the industrial system). In such situations the risks that are associated with each noncompliance will be prioritized, and corrective and/or preventive measures will be taken in order to minimize the risks (see, e.g., Petrescu, 2019).

The concept of sustainable design (https://mec.tuiasi.ro/wp-content/uploads/2020/01/curs-34.pdf) involves multiple approaches, referring to all processes that are specific to the circular economy or only to some of them. The concept of sustainable design can be found in practice in various forms, such as the following:

- *Regenerative design or recyclability*. The key idea of this concept (cradle to cradle, a concept that was launched in 2002 by the architect William McDonough and the chemist Michael Braungart (https://mcdonough.com/cradle-to-cradle/)) is that the end use of a product should not lead to its unusability. Parts of the product or the product as a whole should be reintroduced into the economic circuit, generating inputs in another system. This requires effective life cycle planning and identifying ways to close all possible loops (https://mec.tuiasi.ro/wp-content/uploads/2020/01/curs-34.pdf; Billatos et al., 1997).
- *Biomimetics*. Biomimetics is "the practice of designing materials, processes, or products that are inspired by living organisms or the relationships and systems formed by living organisms." (Asheen, 2009), (https://interfaceinc.scene7.com/is/content/InterfaceInc/Interface/EMEA/WebsiteContentAssets/Documents/25th%20anniversary%20report/English/wc_eu-lessonsforthefuture-en.pdf?utm_source = interface&utm_medium = pdf&utm_campaign = sustainability-en_gb-organic&utm_content = sustainability-report).

TABLE 2.2 Identification of nonconformities (risk analysis) characteristic of industrial systems in the circular economy.

Process or technology	Nonconformities type	Nonconformities effect	Nonconformities causes	Corrective/preventive measures
0	*1*	2	3	*4*
Design	Frequent product replacement and acute demand for new productsImpossibility of disassembling the product	High consumption of natural resourcesLow degree of reuse and recirculationHigh energy consumptionHigh consumption of materials	Short product lifeMonobloc or classic design projectsroducts with a technical service life exceeding their esthetic service lifeOversized products	Sustainable designProduct design to meet the possible future needs of usersProduct design in modular structureOptimizing product life by increasing reliability and durabilityDesign for a minimum production of waste during production
Manufacturing	Products dangerous for consumption and/or unfriendly to the environmentUnfriendly operating conditionsHigh consumption of raw materialsHarmful gaseous, liquid, or solid emissions during productionTechnological lossesHigh energy consumption	Increased scrap rateLow level of safety at workLow customer satisfactionHigh consumption of natural resourcesEnvironmental pollutionLow efficiency of technological processes	se of hazardous, toxic, or environmentally unfriendly materialsInadequate technology (morally degraded or inappropriate for the use of recycled materials)Inefficient energy flowsLeakage in the production flowInadequate raw materialUse of oversized semifinished products	Promoting technological innovationReengineering or modernization of productionelection of raw material (possibly suppliers)Elimination or reduction in the use of hazardous or toxic materials that are not environmentally friendlyOptimization of energy flows (energy recovery)Minimization and recycling of residues and waste from production processes inside or outside the factoryDesign for a minimum production of waste during productionMinimization of the number of materials used in the manufacture of products
Consumption	Large amount of degraded or unused productsLarge quantities of consumables (e.g., batteries, toner, lubricants) usedFluid leaks (liquids, gases)Impossibility of	Increased amount of wasteConsumer dissatisfactionHigh consumption of natural resourcesHigh energy consumptionLoss of	olume of purchases of products (especially food) oversized (unsuitable for actual consumption)Short product lifeProducts with a technical service life exceeding	Correlating the volume of purchases with normal consumption needsUse of repairable or reusable productsReuse of consumablesPeriodic inspection of household installations and equipmentUse of leak detection equipmentEducation and

	disassembling the productsarmful emissionsLarge amount of unusable waste	resourcesImpossibility of repairing the productsEnvironmental pollutionLow degree of reuse or recirculation	their esthetic service lifeLeakage in transmission systems, distribution, use of domestic resourcesNonremovable productsUse of polluting products or technologiesLack of interest in selective waste collection	accountability regarding the need to recycle products
Collection	High share of unusable wasteHigh amounts of contaminated wasteHazardous emissionsFrequent product replacement and acute demand for new productsImpossibility of disassembling the product	Low collection rateCollection of contaminated wasteEnvironmental pollutionHigh energy consumption	Oversized collection networkAbsence of a selective collection programWaste incinerationProducts with a short lifeProducts with a high content of pure or virgin materialsNonremovable products	Investments in selective waste collection networksEducation and accountability in connection with the need to recycle productsUse of repairable or reusable productsProduct design in modular structuresOptimizing product life by increasing reliability and durability
Recovery	Raw materials (resulting from recovery) that are unsuitable for usePoor-quality productsEnvironmental pollutionLarge amounts of wasteExpensive recovery technologies	Low amount of recovered or recycled productsecovered materials of poor quality (low degree of purity)High energy consumptionHigh consumption of natural resources	Low stocks of reusable wasteInadequate technology, lack of recovery or recycling technologyProducts containing heat-resistant polymersOversized productsComplex products containing various materials	Reducing the use of hazardous and toxic materials that are not environmentally friendlyReducing the use of high-energy materials (e.g., aluminum)Use of renewable, recyclable, and/or recycled materialsAvoidance of oversizing products and packagingDesign for a minimum production of waste during productionMinimizing the number of materials used in the manufacture of products

Such an approach is based on the two-way connection between nature and design. Either design seeks to reproduce efficient and/or healthy systems in nature or nature offers challenges for design. Both forms involve looking for environmentally friendly solutions (not necessarily understandable, healthy, or sustainable).

- *Green chemistry*. Also known as sustainable chemistry or environmentally friendly chemistry, green chemistry, as defined by Paul Anastas and John Warner, is intended to be "that redesign of chemical processes that reduces or eliminates the use and generation of hazardous substances" (https://www.scienceinschool.org/wp-content/uploads/2015/02/green_chemistry.pdf; Anastas and Warner, 1998).
- *Design for disassembly*. Design for disassembly, abbreviated DfD, involves the implementation of design concepts that allow easy recovery of parts, components, and materials from products at the end of their life. Recycling and reuse are proving to be sustainable technologies only for products that can be disassembled cleanly and efficiently; otherwise, recycling and reuse may be impossible or may involve prohibitive costs. The ability to reuse products involves the extent to which they retain their value after use (https://www.c2ccertified.org/news/article/green-grows-up-building-in-the-age-of-the-circular-economy).
- *Design for environment*. Design for the environment, abbreviated DfE, aims to reduce pollution and risks that can affect people and the environment. Products that are subject to this principle must be environmentally friendly in all phases and processes that are specific to the circular economy.
- *Product management*. Product management, also known as extended product responsibility, is based on the principle that all those involved in the life cycle of a product must demonstrate responsibility for reducing its impact on the environment. The application of this concept is based on voluntary partnerships between producers, traders, government, and nongovernmental organizations to create efficient waste reduction systems and practices (https://mec.tuiasi.ro/wp-content/uploads/2020/01/curs-34.pdf).
- *Ecological marketing*. Ecological marketing (Billatos et al., 1997) or green marketing (green communication, green consumers, environmentally friendly services, environmental governance, recycling behavior, etc.) refers to the ability to develop high-quality activities that promote the use of environmentally friendly products that are simultaneously in line with consumer expectations (Peattie, 2001; Radnovic, Ilic, & Zivkovic, 2012; Ottman, Stafford, & Hartman, 2006).

As can be seen from the preceding analysis and Table 2.2, design technology has an essential role in guaranteeing the viability of the circularity of consumption. Product characteristics, such as life span, modular design, or sizing, influence most of the other processes in terms of technology or behavior. Similarly, education is a defining element for the implementation of the circular economy, not only for consumers but also for all other processes that are involved in the recovery and recycling of resources.

2.6 Evaluation of the level of reliability in the case of systems related to the circular economy

The notions of reliability, risk, and technical security are extended to systems related to the circular economy in order to assess the level of reliability and identify the vulnerable components.

Reliability and the security attributed to the elements (systems) related to the circular economy are closely related to the notion of failure.

Among the main elements (processes and systems) that compose the circular economy are design, manufacture, distribution, consumption, collection, and recovery. These elements are studied mainly by reliability block diagram (RBD) type analyses and fault tree analyses (FTA) in order to identify how they affect the reliability of the circular economy as a whole (Ilinca, Spireanu, Dumbrava, & Iacob, 2011).

Failure means cessation of the ability of the element (e.g., a component part of a system within the circular economy) to perform its required functions correctly. Failure refers to the appearance and manifestation of nonconformities that are specific to each process or system.

The functioning of the circular economy is based on the hypothesis that the component processes or systems are considered to have a single defect. The renewal processes that take place at their level aim at restoring the initial level of reliability.

If the study of the operating behavior of circular economy systems takes into account only the first failure of the system that appeared after its entry into operation (i.e., the first entry into operation), then the systems in question are called systems without renewal. If the behavior of the system between two consecutive failures is also analyzed, within the limits of the imposed operation duration (until scrapping), the investigated systems are called systems with renewal.

From a qualitative point of view, the concept of the reliability of a system designates the ability of that system to correctly perform its functions imposed, under specified operating conditions, over an established period of time (https://pdfcoffee.com/factorii-care-influenteaza-fiabilitatea-indicatori-de-fiabilitate-pdf-free.html).

The duration of time in which the component elements of the circular economy fulfill their functions is denoted by τ_0 [h,d,an], where τ_0 represents a continuous random variable (Verma and Peterson, 1995; Popescu, 1998):

$$\tau_0 : E \rightarrow [0; +\infty)\tau_0 = \tau_0(e) \tag{2.1}$$

Quantitatively, the concept of reliability of a system (such as design, manufacture, or distribution) designates the probability that the system will correctly perform the functions imposed, under specified operating conditions, over an established period of time. The reliability function, defined on the real-time axis and taking values in the interval [0;1], represents the probability that in the current time interval [0;τ), there is no failure of the system (Popescu, 1998; Ilincă, 2003; Nicolcioiu, 2003):

$$R : [0; +\infty) \rightarrow [0; 1]R(0, \tau) = R(\tau) = P(\tau_0 \geq \tau) \tag{2.2}$$

The unreliability of a process or system represents, from the same quantitative point of view, the probability that in the current time interval [0;τ), the failure of the system will occur (https://pdfcoffee.com/factorii-care-influenteaza-fiabilitatea-indicatori-de-fiabilitate-pdf-free.html; Ilincă, 2003; Nicolcioiu, 2003):

$$F : [0; +\infty) \rightarrow [0; 1]F(0, \tau) = F(\tau) = P(\tau_0 < \tau) \tag{2.3}$$

2.7 Reliability of complex systems and evaluation methods applied to the elements that compose the circular economy

The circular economy can be likened to a set of components that are functionally linked to each other in order to highlight a causal law as a whole.

From the point of view of reliability, the circular economy depends on the level of reliability of the component elements (design, manufacture, distribution, etc.) and of and dependency links between them. If an element fails, the circular economy is more or less affected, depending on the nature of the nonconformity (Verma and Peterson, 1995; Popescu, 1998).

Knowing the level of reliability related to the design, manufacture, and distribution and the realization of a logical scheme (a graphical representation). Knowing the component parts that are interacting, each component is associated with one of the states "component in operation," "defective component," "component in standby," and so on. The reliability model is given by the operating scheme related to the circular

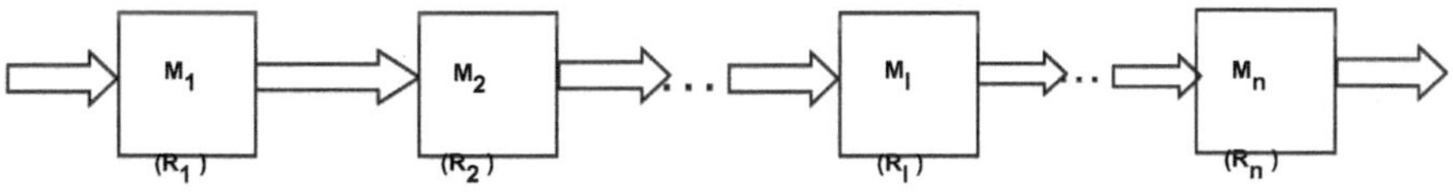

FIGURE 2.9 Series system with n elements.

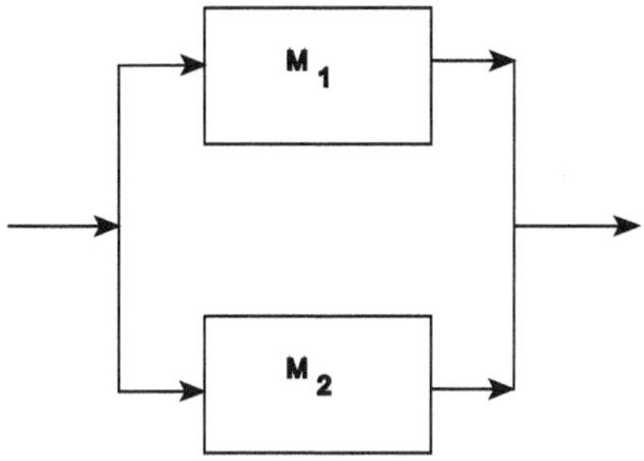

FIGURE 2.10 Parallel system with two elements.

economy. It involves the decomposition of the assembly into component parts that do not depend on each other in terms of failure possibilities.

The series connection (Fig. 2.9) assumes that the failure of any element leads to the failure of the analyzed assembly (Ilinca, Spireanu, Dumbrava, & Iacob, 2011; https://pdfcoffee.com/factorii-care-influenteaza-fiabilitatea-indicatori-de-fiabilitate-pdf-free.html).

The reliability of the circular economy supposes consideration of the component parts, and the good functioning is the result of the behavior of all the parts.

It is considered that the reliability is given by the good operation times related to the component parts, and in the case of the series connection the calculation expression is given by the following relation (http://www.amaniu.ase.ro/studenti/masterCSIE/cap%206.doc):

$$R_s(t) = P[T > t] = P\left[\bigcap_{k=1}^{n} (T > t)\right] = P\prod_{k=1}^{n}[T_k > t] = \prod_{k=1}^{n} R_k(t) \tag{2.4}$$

Obviously, the reliability of an assembly consisting of elements that are connected in series is lower than the lowest of the reliability values of the component elements.

Use of the serial connection is not recommended if the components have very different values in terms of reliability.

The parallel connection assumes that the failure of the entire analysis occurs only if the component parts fail at the same time. The typical connection scheme of the elements in parallel is shown in Fig. 2.10. Compared to the series connection, the reliability of the parallel connection system is higher than the highest of the reliability of the elements that make up the whole.

The unreliability of the parallel connection is given by the following relation (http://www.amaniu.ase.ro/studenti/masterCSIE/cap%206.doc):

$$F_s(t) = P[T < t] = P\left[\bigcap_{k=1}^{n} (T < t)\right] = P\prod_{k=1}^{n}[T_k < t] = \prod_{k=1}^{n} F_k(t) \tag{2.5}$$

Such systems lead to the achievement of a high degree of reliability. Failure of the system is possible when all the component elements fail, but operation of the system is ensured when at least one element operates.

The reliability of the parallel connection is

$$R_s = 1 - \prod_{i=1}^{n}(1 - R_i) \tag{2.6}$$

Starting from the elementary structures of series and parallel type, theoretically, one can analyze the reliability of any system reducible to a combination of such structures. In reality the structure of a system cannot always be reduced to series and parallel combinations. In such cases, determination of the system's reliability is more difficult and involves resorting to the so-called structure function.

In a circular model of industrial systems (Fig. 2.11) that is based on six main components (process type), RBD and FTA techniques may be applied to identify the impact of the reliability of process components on the overall reliability of the circular economy.

FTA has the role of highlighting the elements that have an important contribution in evaluating an unwanted event that may occur during the life of the investigated system. It is applied in reliability, security, and risk analyses (https://pdfcoffee.com/factorii-care-influenteaza-fiabilitatea-indicatori-de-fiabilitate-pdf-free.html; Popescu, 1998; Nicolcioiu, 2003).

FTA, as an analysis technique applied to the circular economy, involves knowing the particularities of the circular economy. For application an unwanted event is established in an attempt to highlight all the elements that can lead to it. The unwanted event is called the top event (SAE Aerospace Recommended Practice ARP5580, 2001; IEC 61511 -Functional safety —Safety instrumented systems for the process industry sector —Part 1 to 3, 2003).

Carrying out the analysis requires that immediately after the top event, the independent, basic events that can generate it be taken into account. Next, each of them develops, also through deductive analysis, in other events that can generate them, and so on. Going from close to close, in the end, it reaches the primary event, which can contribute to the generation of the chain of unwanted events (failures) to the top event (Nicolcioiu, 2003).

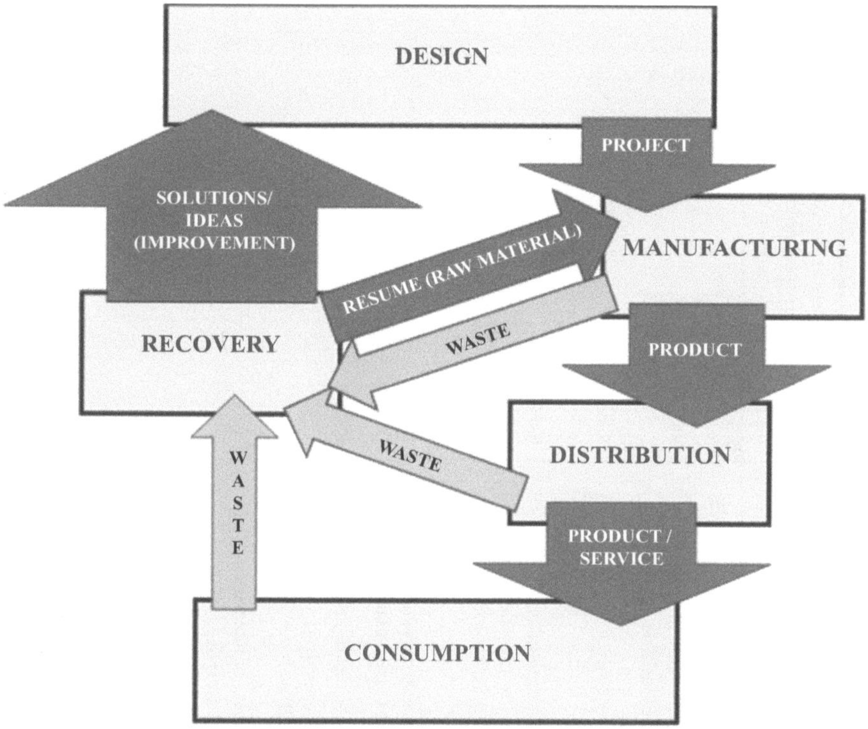

FIGURE 2.11 The circular model of industrial systems (process map).

The graphical representation related to the analysis resembles a tree, an aspect that led to the name logical fault tree analysis.

The reliability study of the circular economy (by RBD or FTA analysis) is performed starting, mainly from the afferent process by highlighting the type of existing nonconformities and by specifying the causes of the nonconformities (Table 2.3, columns 0, 2, and 4).

The connection between the events related to the FTA analysis is made through the AND and OR type logic gates.

The logic tree that is necessary to evaluate the reliability of the cyclic economy (based on the elements presented in Table 2.3) in the RBD-type scheme are presented in Figs. 2.12–2.17.

The input data are found (in Table 2.3, column 5) in the form of the weight of the causes in the occurrence of nonconformities.

In essence, the weight (expressed as a percentage) represents the value of the unreliability related to the nonconformities, and when the complementarity relationship between reliability and unreliability are taken into account, the reliability automatically results.

TABLE 2.3 The elements component (process type) related to the circular economy and the related nonconformities.

Process	Type of nonconformities	Causes of nonconformities	The weight value of the causes in the occurrence of nonconformities (probable values based on the authors' experience, used to exemplify the method) (%)
0	2	4	5
Design	Frequent replacement of products	Short product lifeProducts with a technical service life exceeding the esthetic service lifeMonobloc projects	602020
	Impossibility to disassemble the product	Oversized productsMonobloc projects	5050
Manufacturing	Dangerous productsFor consumption and/or unfriendly to the environment	Use of hazardous, toxic, or environmentally materialsInadequate technology (morally degraded or inadequate to use recycled materials)Inadequate raw material	403030
	High energy consumption	Inefficient energy flowsInadequate technology (morally degraded or inadequate to use recycled materials)	050
	Technological losses	Inadequate technology (morally degraded or inadequate to use recycled materials)Use of oversized semifinished productsInadequate raw material	404020
	Harmful emissions during manufacture	Leakage in the production flowUse of hazardous, toxic, or environmentally unfriendly materials	5050
	High consumption of raw materials	Inadequate raw materialInadequate technology (morally degraded or inadequate to use recycled materials)Use of oversized semifinished products	303040
	Improper operating conditions	Inefficient energy flowsInadequate technology (morally degraded or inadequate to use recycled materials)	5050

Distribution	Unsold product	Long-haul transport routesInadequate marketingUse of inappropriate packaging (e.g., PVC, aluminum)	157015
	High energy consumption	Incorrect market estimateIndividual packaging of productsExcess stocks(above market assimilation capacity)	202060
	Toxic emissions	Use of inappropriate packaging (e.g., PVC, aluminum) Excess stocks(above market assimilation capacity)	5050
	Large amount of packaging	se of inappropriate packaging (e.g., PVC, aluminum) Individual packaging of productsCustom packaging	404020
	High consumption of environmentally hazardous materials	Use of inappropriate packaging (e.g., PVC, aluminum) Individual packaging of products	5050
Use	Degraded products	Oversized procurement volume(especially food) unsuitable for actual consumptionhort product lifeLeakage in transport systems, distribution, use of domestic resources	403030
	Large quantities of consumables (batteries, toner, lubricants, etc.)	Oversized procurement volumeDisinterest in selective waste collection	4060
	Harmful emissions	Leakage in transport systems, distribution, use of domestic resourcesShort product life	7030
	Unusable waste	Oversized procurement volumeDisinterest in selective waste collection	5050
Collection	High amounts of contaminated waste	Oversized collection networkAbsence of a selective collection program	5050

(*Continued*)

TABLE 2.3 (Continued)

Process	Type of nonconformities	Causes of nonconformities	The weight value of the causes in the occurrence of nonconformities (probable values based on the authors' experience, used to exemplify the method) (%)
	Hazardous emissions	Waste incinerationAbsence of a selective collection program	6040
Recovery	Raw materials (resulting from recovery) unsuitable for use	Low stocks of reusable wasteProducts containing heat-resistant polymersComplex products containing various materials	204040
	Environmental pollution	Inadequate technology, lack of recovery or recycling technologyProducts containing heat-resistant polymersComplex products containing various materials	204040
	Large amounts of waste	Oversized productsLow stocks of reusable waste	5050
	Expensive recovery technologies	Inadequate technology, lack of recovery or recycling technologyComplex products containing various materials	5050

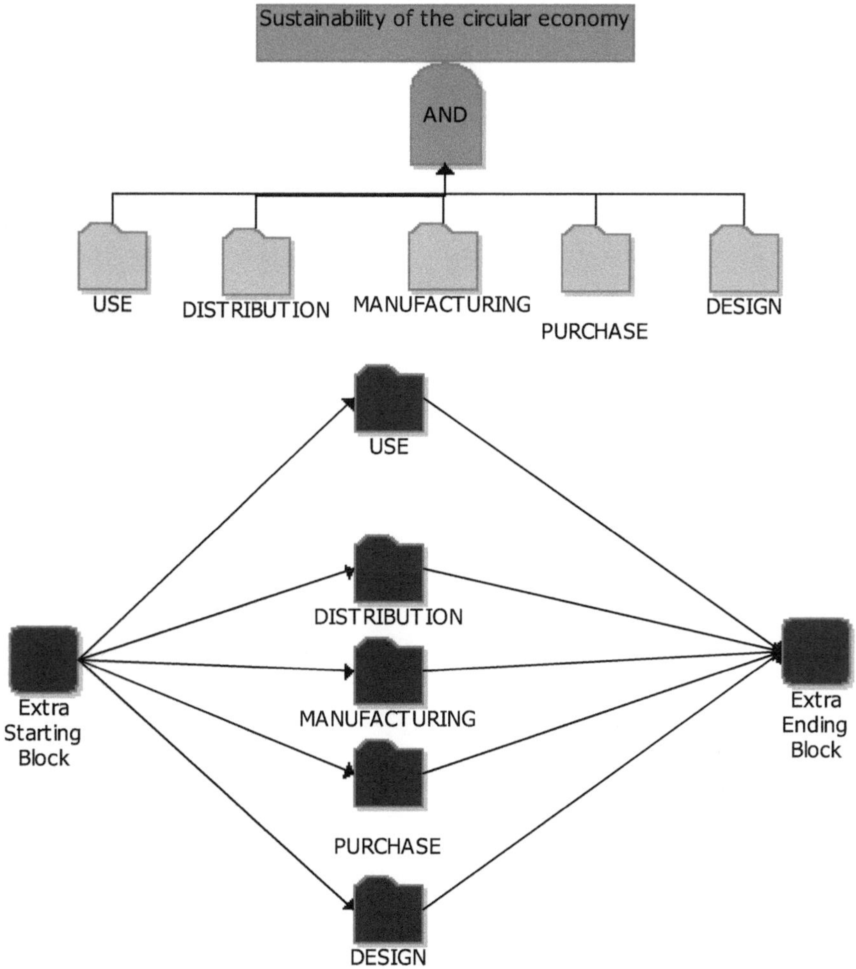

FIGURE 2.12 The logic tree and the RBD-type scheme of the reliability study related to the circular economy.

For example, in the case of ***DESIGN*** (see Fig. 2.13), the nonconformity of "frequent replacement of products" has as established causes: products with a technical life span that exceeds their esthetic life span (20%), reduced life span of product (80%), and monobloc projects (20%). Starting from these hypotheses, the values of reliability (R) automatically result, as follows:

- Products with a technical service life exceeding their esthetic service life, $R = 0.8$.
- Reduced product life, $R = 0.2$.
- Monobloc projects, $R = 0.2$.

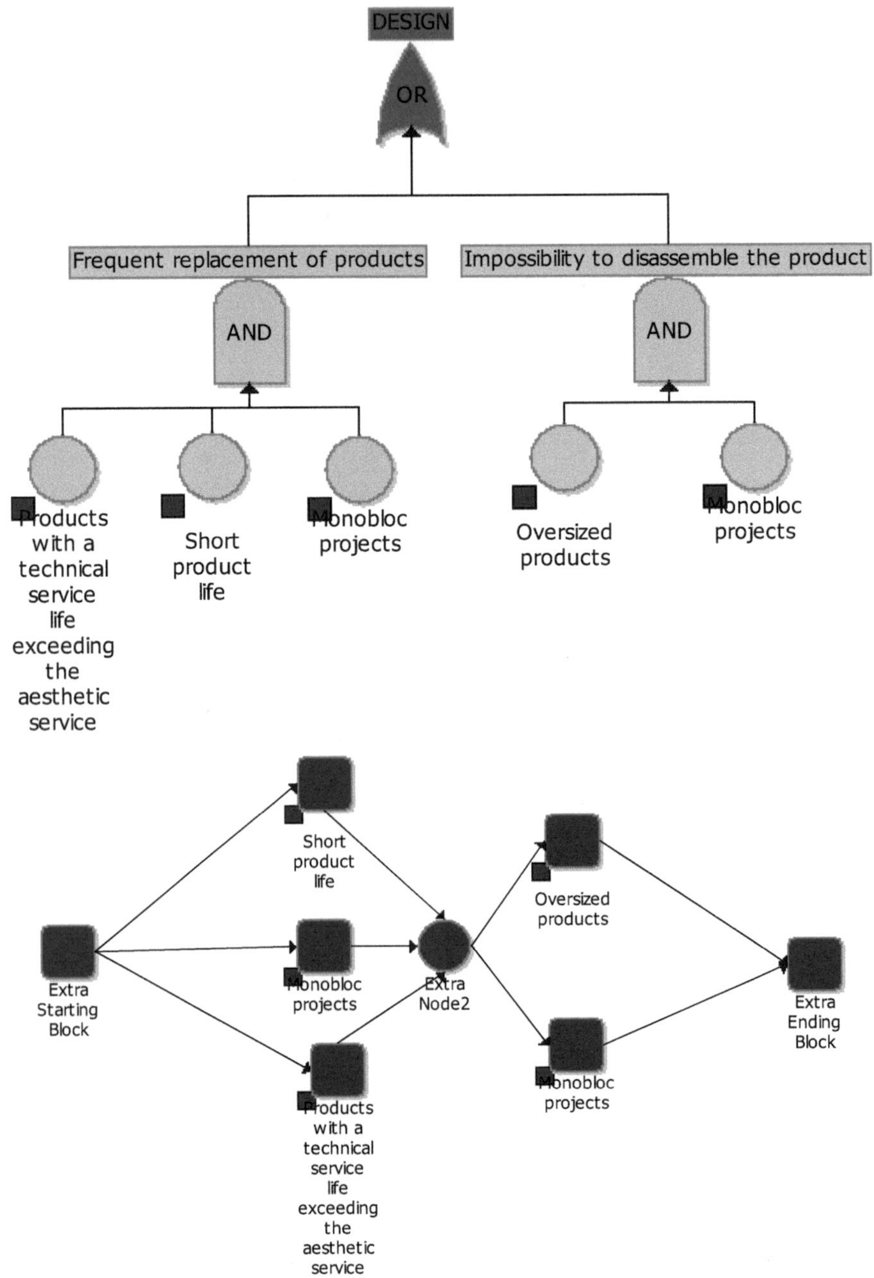

FIGURE 2.13 The logic tree and the RBD-type scheme of the reliability study related to ***DESIGN*** as a component part of the circular economy (see Table 2.3).

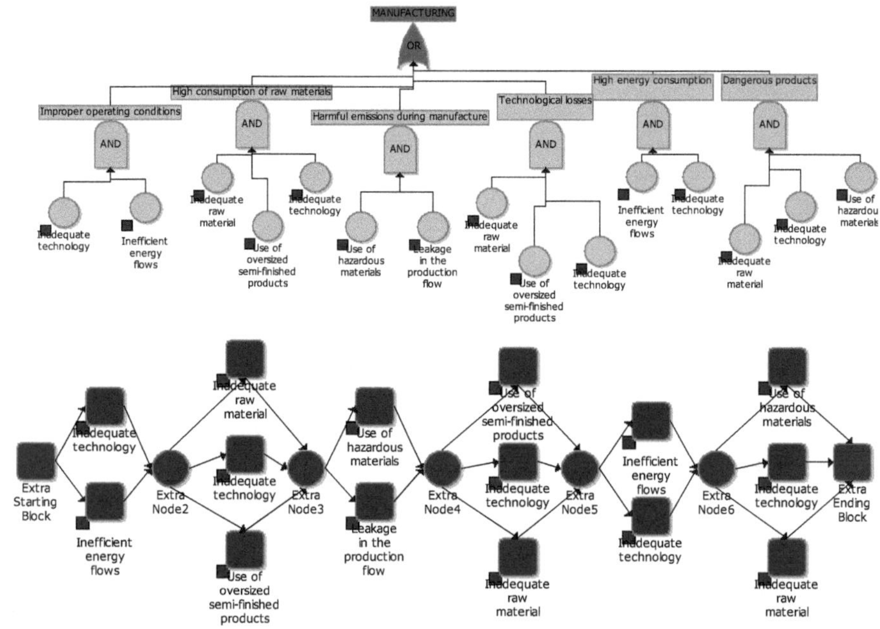

FIGURE 2.14 The logic tree and the RBD-type scheme of the reliability study related to ***MANUFACTURING*** as a component part of the circular economy (see Table 2.3).

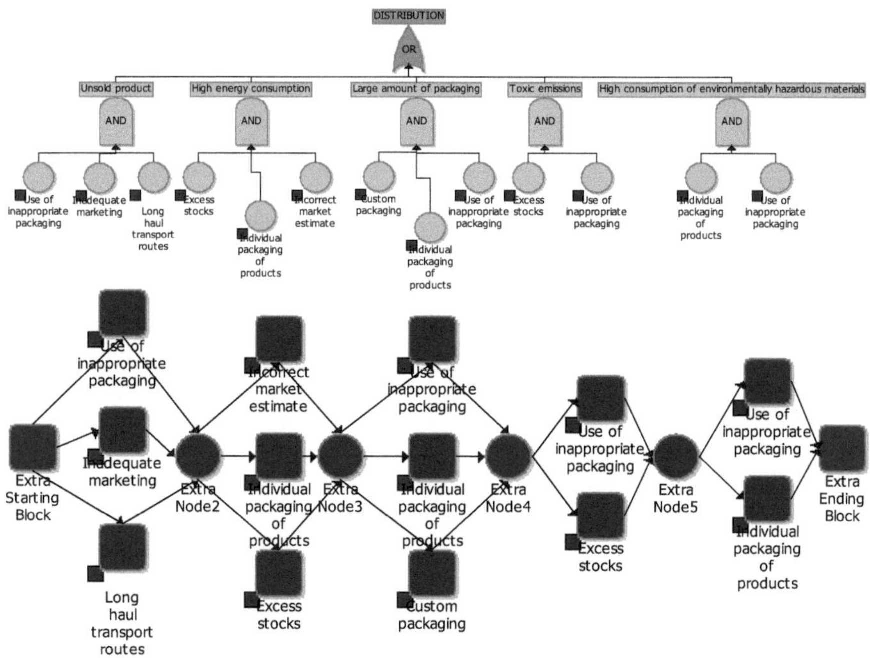

FIGURE 2.15 The logic tree and the RBD-type scheme of the reliability study related to ***DISTRIBUTION*** as a component part of the circular economy (see Table 2.3).

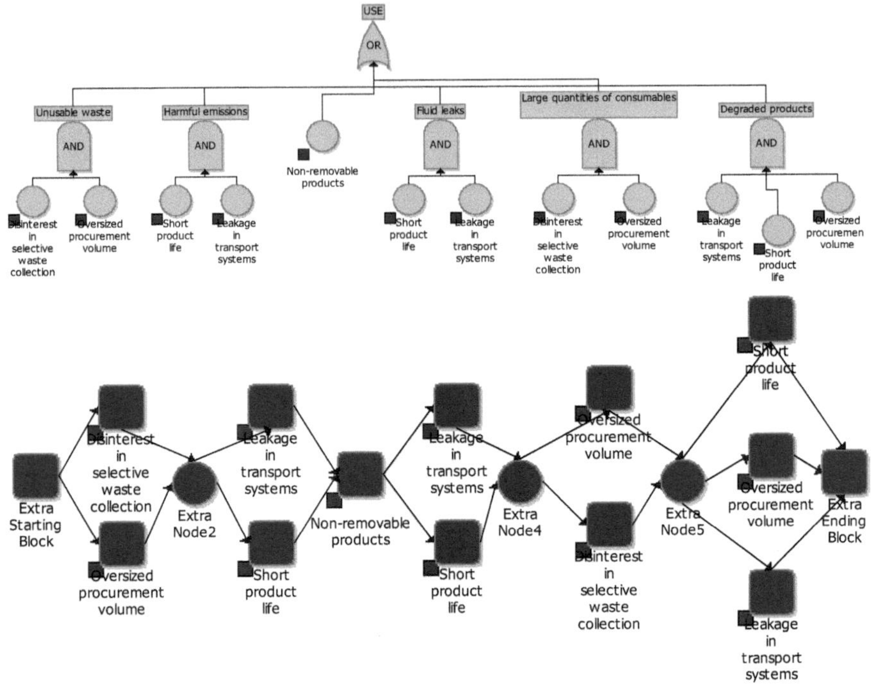

FIGURE 2.16 The logic tree and the RBD-type scheme of the reliability study related to ***USE*** as a component part of the circular economy (see Table 2.3).

Starting from these data, the logical failure tree is solved, the RBD type analysis is performed for each process, and finally the reliability of the circular economy is analyzed (see Fig. 2.9), considering that the process elements are considered as connected in parallel. If we consider this type of connection (parallel), we can see very clearly the influence of each process (design, manufacture, distribution, consumption, collection) on the efficiency of the circular economy. Obviously, the decrease in functionality (reliability) of one of the components of the circular economy does not automatically lead to interruption of the functionality but leads to a decrease in the efficiency.

In the case of ***DESIGN*** (see Fig. 2.13) as a process, the expression of reliability (R) is given by the following relation:

((RExtra Starting Block.RExtra Node2(−2RShort product life. RMonobloc projects + Short product life.RMonobloc projects^2 + RShort product life + RMonobloc projects − RMonobloc projects^2))(RExtra Ending Block(2ROversized products − R Oversized products^2)))

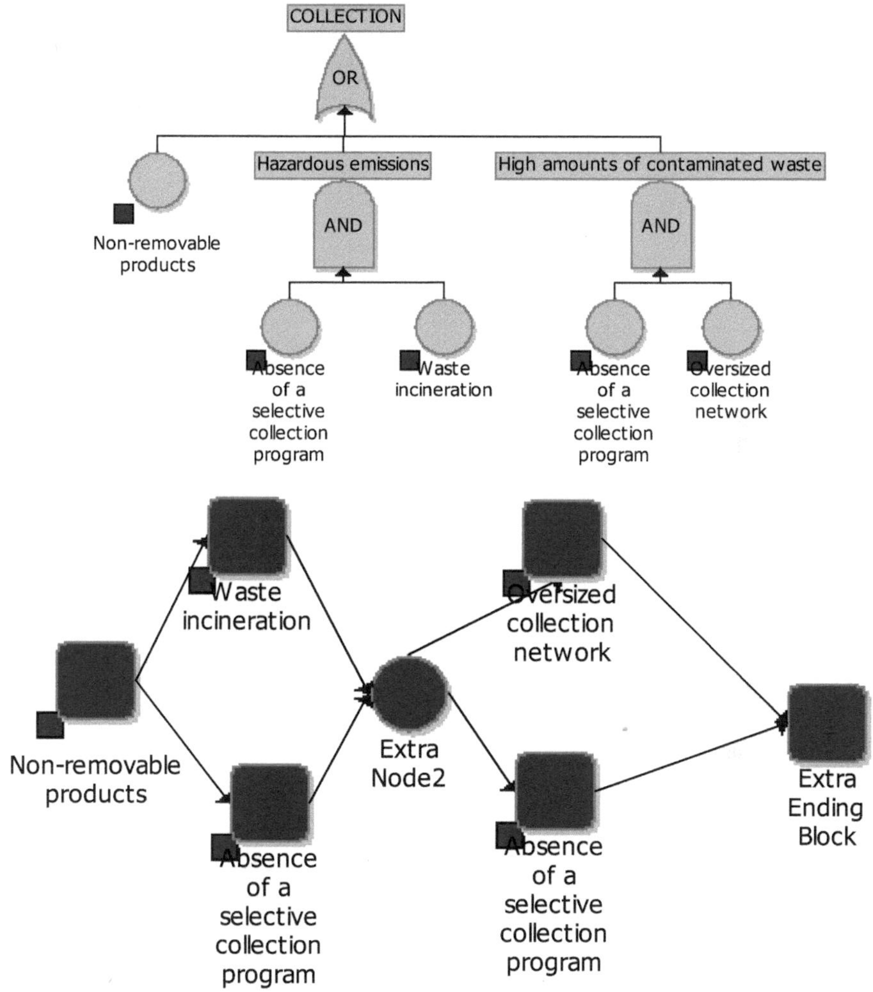

FIGURE 2.17 The logic tree and the RBD-type scheme of the reliability study related to ***COLLECTION*** as a component part of the circular economy (see Table 2.3).

where ROversized products = RMonobloc projects, RMonobloc projects = RProducts with a technical service life exceeding their esthetic service life (Table 2.4).

In the case of simulations, results can be obtained depending on the number of events that are required to exit the parallel joints. In the case of the present simulations, one out of *n* events was considered. Following the evaluations, the reliability values related to the circular economy processes are presented in Tables 2.5–2.9.

TABLE 2.4 The value of reliability related to ***DESIGN***, as a process of the circular economy.

Causes of nonconformities	Reliability	Unreliability
Short product life	0.4	0.6
Monobloc projects	0.8	0.2
Oversized products	0.5	0.5
onobloc projects	0.5	0.5
Products with a technical service life exceeding their esthetic service life	0.8	0.2
Reliability =	0.732	

TABLE 2.5 The value of reliability related to ***MANUFACTURING*** as a process of the circular economy.

Causes of nonconformities	Reliability	Unreliability
Use of hazardous materials	0.6	0.4
Inadequate technology	0.7	0.3
Inadequate raw material	0.7	0.3
Inefficient energy flows	0.5	0.5
Inadequate technology	0.5	0.5
Use of oversized semifinished products	0.6	0.4
Inadequate technology	0.6	0.4
Inadequate raw material	0.8	0.2
Use of hazardous materials	0.5	0.5
Leakage in the production flow	0.5	0.5
Inadequate raw material	0.7	0.3
Inadequate technology	0.7	0.3
Use of oversized semifinished products	0.6	0.4
Inadequate technology	0.5	0.5
Inefficient energy flows	0.5	0.5
Reliability =	0.3795	

By using the results of the simulations obtained for each process, based on the RBD or FTA type diagram in Fig. 2.12, the necessary information can be obtained regarding the reliability of the circular economy.

TABLE 2.6 The value of reliability related to ***DISTRIBUTION*** as a process of the circular economy.

Causes of nonconformities	Reliability	Unreliability
Use of inappropriate packaging	0.5	0.5
Individual packaging of products	0.5	0.5
Use of inappropriate packaging	0.4	0.6
Excess stocks	0.6	0.4
Individual packaging of products	0.6	0.4
Use of inappropriate packaging	0.6	0.4
Excess stocks	0.4	0.6
Individual packaging of products	0.8	0.2
Incorrect market estimate	0.8	0.2
Inadequate marketing	0.3	0.7
Long-haul transport routes	0.85	0.15
Use of inappropriate packaging	0.85	0.15
Custom packaging	0.8	0.2
Reliability =	0.53	

TABLE 2.7 The value of reliability related to **COLLECTION**, as a process of the circular economy.

Causes of nonconformities	Reliability	Unreliability
Oversized collection network	0.5	0.5
Absence of a selective collection program	0.5	0.5
Waste incineration	0.4	0.6
Absence of a selective collection program	0.6	0.4
Nonremovable products	0.8	0.2
Reliability =	0.456	

The mathematical expression of reliability related to the circular economy (based on the logical scheme from Fig. 2.12) is given as follows:

(RExtra Ending Block(RExtra Starting Block.RNew Node (RUSE.RDISTRIBUTION.RMANUFACTURING.RCOLLECTION. RDESIGN-RUSE.RDISTRIBUTION.RMANUFACTURING. RCOLLECTION-RUSE.RDISTRIBUTION.RMANUFACTURING. RDESIGN-RUSE.RDISTRIBUTION.RCOLLECTION.RDESIGN-RUSE.

TABLE 2.8 The value of reliability related to ***USE***, as a process of the circular economy.

Causes of nonconformities	Reliability	Unreliability
Leakage in transport systems	0.7	0.3
Short product life	0.7	0.3
Oversized procurement volume	0.6	0.4
Oversized procurement volume	0.6	0.4
Disinterest in selective waste collection	0.4	0.6
Leakage in transport systems	0.3	0.7
Short product life	0.7	0.3
Leakage in transport systems	0.3	0.7
Short product life	0.7	0.3
Oversized procurement volume	0.5	0.5
Disinterest in selective waste collection	0.5	0.5
Nonremovable products	0.7	0.3
Reliability =	0.2401	

TABLE 2.9 The value of reliability related to ***the circular economy***.

Causes of nonconformities	Reliability	Unreliability
USE	0.2401	0.7599
DISTRIBUTION	0.53	0.47
MANUFACTURING	0.3795	0.6205
COLLECTION	0.456	0.544
DESIGN	0.732	0.268
Reliability =	0.9677	

RMANUFACTURING.RCOLLECTION.RDESIGN-RDISTRIBUTION. RMANUFACTURING.RCOLLECTION.RDESIGN + RUSE. RDISTRIBUTION.RMANUFACTURING + RUSE.RDISTRIBUTION. RCOLLECTION + RUSE.RDISTRIBUTION.RDESIGN + RUSE. RMANUFACTURING.RCOLLECTION + RUSE.RMANUFACTURING. RDESIGN + RUSE.RCOLLECTION.RDESIGN + RDISTRIBUTION. RFABRICATION.RCOLLECTION + RDISTRIBUTION. RMANUFACTURING.RDESIGN + RDISTRIBUTION.RCOLLECTION. RDESIGN + RMANUFACTURING.RCOLLECTION.RDESIGN-RUSE.

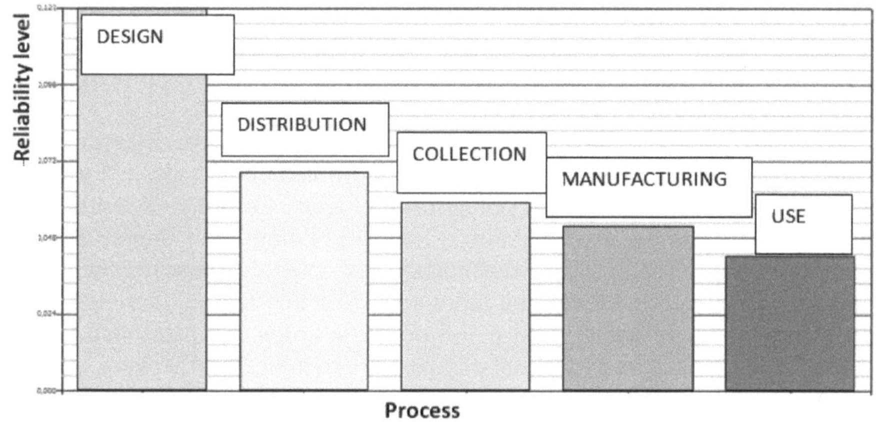

FIGURE 2.18 The importance of each process for the level of reliability related to the circular economy.

RDISTRIBUTION-RUSE.RMANUFACTURING-RUSE.RCOLLECTION-RCUSE.RDESIGN-RDISTRIBUTION.RMANUFACTURING-RDISTRIBUTION.RCOLLECTION-RDISTRIBUTION.RDESIGN-RMANUFACTURING.RCOLLECTION-RMANUFACTURING.RDESIGN-RCOLLECTION.RDESIGN + RUSE + RDISTRIBUTION + RMANUFACTURING + RCOLLECTION + RDESIGN)))

The importance of each process for the level of reliability related to the circular economy can be seen in Fig. 2.18.

The analyses that are performed to evaluate the level of reliability related to the circular economy highlight a global model, which is a model that can be used on a large scale. The input data proposed by the authors (Table 2.3, column 5) will be replaced with data obtained from monitoring the processes involved.

The main purpose of the analysis was to evaluate the reliability and determine the weight that each process as a whole has in the reliability of the circular economy. Obviously, the processes with a relatively low level of reliability can be improved by acting on the causes of nonconformities, reducing their number and/or their frequency of occurrence. Future studies proposed by the authors refer to the possibility of identifying the law of distribution of the causes of nonconformities (Table 2.3, column 4).

2.8 Conclusion

In the current economic circumstances, consumption must be analyzed as a cumulative action of all phases and technologies that are found in the manufacturing and use cycle of a product, potentially generating waste.

Smart consumption refers to the application of selective and differentiated consumption technologies.

A functional and efficient circular economy requires the integration of innovative manufacturing strategies and optimized consumption technologies coupled with efficient recycling technologies. Compliance with these conditions ensures the sustainability of each technology and the sustainability of the entire cycle. The sustainability of the company (products or manufacturing) is conditioned by innovation. For this reason, there is a need to encourage innovation, a culture of innovation at the organizational level.

The innovation strategy must not be subordinated to the idea of market capture. In establishing the innovation strategy, it is necessary to correctly understand the relationships between the inputs (resources, from a material point of view, and requirements, if we refer to the customer or consumer) and the outputs from the organization (products and services, from material point of view, and level of satisfaction, if we refer to the customer). The place and role of the consumer (customer) in defining the company's strategy in general and the innovation strategy in particular must be understood.

In a circular economy, any new product or service must be designed on the basis of the circularity principle. Increasing the level of cooperation in and between manufacturing processes and supply chains as well as between traders and customers can reduce costs, waste, and harmful effects on the environment.

An analysis of the reliability of a circular economy system must start from the identification of nonconformities that may occur during the implementation of the processes that contribute to the circularity of the use of resources.

The processes of the economic circuit are naturally interdependent, not only because of the supplier-beneficiary relationship but as a result of identifying common causes in the manifestation of nonconformities.

The reliability of the circular economy depends on the elements that can have a major impact on the system as a whole. FTA was the main method used in this chapter for assessing the reliability of the circular economy. The proposed methodology can be applied to any real economic system, and it will be based on real data obtained from prior monitoring.

References

Anastas, P. T., & Warner, C. (1998). *Green chemistry theory and practice.* New York: Oxford University Press.

Asheen, P. (2009). *"Biomimicry," Berkshire encyclopedia of sustainability: sustainable development.* New York: Berkshire Publishing Group.

Billatos, I., Basaly A., S., & N. (1997). *Green technology and design for the environment*. Taylor & Francis.

IEC 61511 -Functional safety –Safety instrumented systems for the process industry sector –Part 1 to 3, 2003.

Ilincă, C. (2003). *Metode fundamentale de evaluare a riscurilor tehnice/tehnologice*. Curs Phare, Ploiesti.

Ilinca, C. N., Spireanu, G. H., Dumbrava, V., & Iacob, S. (2011). *Identificarea şi calculul nivelului de integritate a securităţii SIL în domeniul energetic* (pp. 160–pages). Ploieşti: Editura Universitatii Petrol-Gaze din Ploiesti.

Nicolcioiu A. (2003). *Strategiiprivindingineriasimanagementulrisculuitehnic/tehnologic in procesarea dehidrocarburi* (Ph.D. thesis), Ploiesti.

Ottman, A. J., Stafford, R. E., & Hartman, L. C. (2006). Avoiding green marketing myopia ways to improve consumer appeal for environmentally preferable products. *Environment: Science and Policy for Sustainable Development*, *48*(5), 22–36. Available from http://doi.org/10.3200/ENVT.48.5.22-36.

Peattie, K. (2001). Towards sustainability: The third age of green marketing. *The Marketing Review*, 2(2), 129–146.

Petrescu M.G. (2019). *Managementul tehnologiilor industriale*, Editura Universităţii Petrol-Gaze din Ploieşti, Ploieşti.

Popescu D. (1998) *Pavel A.- Risctehnic/tehnologic. Inginerieşi management-Metoda MADS-MOSAR*. Bucureşti, EdituraBriliant.

Radnovic, B., Ilic, M., & Zivkovic, Z. D. (2012). Green marketing and sustainable development-experiences from Republic of Serbia. *Journal of Economic Development, Environment and People*, *1*(3), 77–91.

SAE Aerospace Recommended Practice ARP5580: Recommended failure modes and effects analysis (FMEA) practices for non-automobile applications. SAEInternational, Warrendale, PA, 2001.

Verma, D., & Peterson, E. L. (1995). *Maintainability: A key to effective serviceability and maintenance management*. New York: John Wiley & Sons, Inc.

CHAPTER

3

Energy innovations and sustainable development of circumpolar territories in Russia

Vasilii Erokhin and Gao Tianming
Harbin Engineering University, Harbin, P.R. China

3.1 Introduction

The contemporary situation in the energy sector can be described as a new dual reality. On the one hand, the need for a constant increase in economic growth rates and accelerated development of high-tech energy-intensive industries requires an ever-growing amount of energy generation by all possible means. According to the United Nations (2021a), all people and businesses worldwide should get universal access to affordable, reliable, sustainable, and modern energy and related energy services. On the other hand, the carbon footprint of anthropogenic activity has become so evident that immediate international actions are required to reduce emissions and decarbonize the global energy sector (Panait, Voica, & Radulescu, 2019; United Nations Framework Convention on Climate Change, 2018). In 2020 the average annual concentration of carbon dioxide in the atmosphere reached the ever-high 415 parts per million—an unprecedented event in the earth's geological history (Klimenko, Klimenko, Tereshin, & Mikushina, 2021). The energy sector is blamed for being the main contributor to climate change, with about 73% of all human-caused greenhouse gases (United Nations, 2021a). One of the 17 United Nations Sustainable Development Goals (United Nations, 2021b) calls for urgent action to combat climate change and its impacts. A growing number of scholars (Andrei, Avram,

Innovation and Sustainable Manufacturing
DOI: https://doi.org/10.1016/B978-0-12-819513-0.00001-9

Băncescu, Gâf Deac, & Gheorghe, 2021; Bansal, Zobaa, & Saket, 2005; Khambalkar, Katkhede, Dahatonde, Korpe, & Nage, 2010; Lundy, 2019; Sayigh, 2021) acknowledge that the progression of climate change can be slowed down only by energy conservation, increasing energy efficiency, and reducing or eliminating the use of fossil fuels. In an attempt to curb carbon dioxide emissions and tackle climate change, developed and most developing countries have been revising their energy strategies to pave the way for the green transition of their economies (Kiushkina et al., 2020; Kokorin, 2017; Nosko, 2017; Van, 2014). Having concluded the Paris Agreement on Climate Change in 2015, signatory parties (192 countries and territories) agreed to "substantially reduce global greenhouse gas emissions to limit the global temperature increase in this century to 2°C while pursuing efforts to limit the increase even further to 1.5°C" (United Nations, 2015).

Such ambitious goals require a radical transformation not only of the energy sector itself, but also of the entire global system of economic and social relations—from the use of natural resources to consumption of final goods and services. To keep the climate warming rate below 1.5°C, global carbon dioxide emissions must be reduced by 45% by 2030 and fall to zero around 2050 (United Nations Development Programme, 2021). This threshold can be achieved if two critical conditions are met: concerted efforts of the entire global community and a qualitatively higher level of energy-related technologies. The role of each country is invaluable. Nevertheless, the energy policies of the world's largest developed economies obviously form global energy production and consumption patterns. Among these trend setters is Russia, one of the largest suppliers of various kinds of energy resources (oil, gas, and coal) to the world market. Russia's substantial industrial sector accounts for its significant negative contribution to the total global emissions of carbon dioxide and greenhouse gases into the atmosphere, water and land pollution, and destruction of ecosystems. Russia's attitude to the international climate and energy agenda is all the more important because the bulk of the extraction of hydrocarbons and coal is accounted for the Arctic zone of the country, Siberia, and the Far East. Most contemporary studies (Cai et al., 2021; Luoto, Ojala, & Zajaczkowski, 2019; McCrystall, Stroeve, Serreze, Forbes, & Screen, 2021; Polvani, Previdi, England, Chiodo, & Smith, 2020; Zhilina, 2021) show that the northern hemisphere, particularly the Arctic, is warming significantly faster than the rest of the planet. The region is experiencing unprecedented environmental change. Its effects on fragile northern ecosystems are far more harmful in comparison to those in faster-recovering temperate zones. Arctic warming not only destroys local ecosystems, but also accelerates global climate change. In this light, the duality of the new energy-climate reality is particularly severe for Russia. A continuous increase in the exploitation of energy resources in the Arctic is vital to sustaining economic growth and ensuring the country's energy security. Meanwhile, a decline in output and conservation of Arctic energy resources are both critical for mitigating global climate-related threats.

The solution lies somewhere in between the two extremes. In the case of the Arctic zone of Russia this chapter seeks to develop an understanding of how the "produce more, pollute less" dilemma can be addressed within the framework of a sustainable spatial development approach. There is a chance that sustainable development goals in the High North can be accomplished through the united efforts of Arctic and non-Arctic countries and through the development of appropriate low-emission and low-carbon energy technologies. Thus the chapter aims to show to what extent, in what ways, and what particular energy-related innovations can contribute to the implementation of sustainable development and spatial development goals in circumpolar territories in Russia.

3.2 Russia: energy sector outlook

Russia's fuel and energy complex comprises oil, gas, coal, and peat industries; electric power industry; nuclear and renewable energy sectors; and heat supply system. The share of the fuel and energy complex in investments in fixed assets is 30% of the total volume (40% of the federal budget revenues and over 50% of total exports). Fuel and electricity production exceeds domestic consumption by 90% for oil, 80% for coal, and 50% for natural gas (Table 3.1).

Russia ranks among the world's top countries in hydrocarbon reserves and production and export of energy (in various forms) and energy-related technologies (particularly in the oil, gas, and nuclear sectors). Russia's energy mix is one of the most environmentally friendly (low-carbon) among developed economies. Natural gas accounts for about half of total electricity generation, while nuclear energy, hydropower, and other renewable energy sources account for about one-third.

3.2.1 Oil

Over the past decade, oil production has increased by more than 14%, owing to the exploration of oil fields in Eastern Siberia (Vankorsk cluster) and the Republic of Sakha (Yakutia) as well as in the Caspian Sea. Development of the Prirazlomnoye field has started on the continental shelf in the Arctic Ocean. In 2016 Russian oil and gas tycoon Gazpromneft launched the Arctic Gate offshore oil-loading terminal in the Ob Bay. Oil production in Eastern Siberia and the Far East has increased by a factor of five from 14.3 million tons to 74.6 million tons. In

TABLE 3.1 Russia's projected energy mix, 2018–35.

Parameter	2018	2024		2035		2035–18, Change (%)	
		Low-growth scenario	High-growth scenario	Low-growth scenario	High-growth scenario	Low-growth scenario	High-growth scenario
Oil, million tons							
Production	555.7	556.0	560.0	490.0	555.0	−11.8	−0.1
Domestic consumption	57.3	54.1	52.7	21.6	43.0	−62.3	−25.0
Export	260.6	267.2	269.2	243.7	251.9	−6.5	−3.3
Import	0.5	0.7	0.7	0.7	0.7	+40.0	+40.0
Natural gas (billion m^3)							
Production	727.6	795.1	820.6	859.7	1,000.7	+18.2	+37.5
Domestic consumption	494.2	496.2	509.9	505.3	520.1	+2.2	+5.2
Export, pipeline	220.6	243.9	250.4	255.4	300.6	+15.8	+36.3
Export, LNG	26.9	59.8	65.1	108.0	189.0	+301.5	+602.6
Import	9.4	4.8	4.8	9.0	9.0	−4.3	−4.3
Coal (million tons)							
Production	439.3	448.0	530.0	485.0	668.0	+10.4	+52.1
Domestic consumption	205.3	195.2	213.4	184.0	211.0	−10.4	+2.8
Export	210.3	219.0	272.0	257.0	392.0	+22.2	+86.4
Import	24.3	20.0	19.0	14.5	15.0	−40.3	−38.3
Electricity (billion kWh)							
Production	399.4	389.5	405.2	434.5	475.4	+8.8	+19.0
Domestic consumption	383.8	375.4	391.1	425.7	465.6	+10.9	+21.3
Export	20.5	15.3	15.3	9.9	10.9	−51.7	−46.9
Import	5.0	1.1	1.1	1.1	1.1	−77.8	−77.8

Authors' development based on Government of the Russian Federation. (2020b). Decree #1523-r from June 9, 2020, Energy strategy of the Russian Federation for the period until 2035. *<https://minenergo.gov.ru/node/1026>; Accessed 08.12.21.*

2018 the continental shelf projects produced 29.1 million tons of oil (two times more than in 2008). The extraction of hard-to-recover oil reserves has grown by 6 million tons to 38 million tons. Crude oil exports in 2018 increased by 7.2% compared to 2008, mainly owing to supplies to China and other countries in Asia. The major problem that Russian oil companies face is increased production cost due to the high portion of hard-to-recover reserves in their portfolios amid the overexploitation and growing depletion of mature fields. This complicates the maintenance of stable output volume and necessitates the adoption of more expensive production technologies and innovations. Physical and chemical characteristics of oil have been deteriorating (e.g., an increase in the density and sulfur content). Degrading crude oil quality requires the advancement of oil refinery technologies and substantially increases processing costs.

3.2.2 Natural gas

Natural gas production has increased by 9.1% compared to 2008. Large gas fields are being developed on the Yamal Peninsula (Bovanenkovo, Kharasavey, and South-Tambey), Eastern Siberia (Kovyktinsk, Vankorsk), the Republic of Sakha (Yakutia) (Chayandinsk, Talakansk), and Irkutsk Oblast (Verkhnechonsk). Gazprom exploits deep-lying strata and deposits in the Zapolyarny and Urengoysk fields. An innovative project to extract methane from coal seams is being implemented in Kuzbass (the Kuznetsk Basin in Kemerovo Oblast). Underwater production complexes are used on the shelf of the Sea of Okhotsk to explore and extract oil and natural gas. A booming segment of the natural gas sector is liquefied natural gas (LNG) on the Yamal and Gydan peninsulas in the Arctic. Yamal LNG is based on the feedstock resources of the South-Tambey field with proved and probable reserves estimated at 926 billion m^3 of natural gas and 30 million metric tons of liquid hydrocarbons. The field's production potential amounts to approximately 27 billion m^3 of natural gas per annum (Novatek, 2021b). Arctic LNG-2 will be built in the Gydan Peninsula (Utrenneye field) and will include three LNG trains with a capacity of 6.6 million tons per annum. The total projected capacity of Arctic LNG-2 is 19.8 million tons per annum (Novatek, 2021a). Novatek is about to launch line 4 of Yamal LNG (0.95 million tons) using the new Arctic Cascade gas liquefaction technology. As part of the Yamal LNG project, a transport infrastructure has been created, including a seaport and an airport in Sabetta, Yamal-Nenets Autonomous District. However, there are challenges in the gas sector in the Arctic, somewhat similar to those in the oil sector. Production and transportation costs grow as a result of the depletion of highly productive and shallow reserves, the shift to the development of lower-capacity gas fields located in difficult geological landscapes, and the remoteness of new production sites from major gas consumption centers and transshipment hubs.

3.2.3 Coal

Coal production in Russia increased by about 40% in 2020 compared to 2008, owing to the intensification of open-pit mining (over 75% of the total coal output). The processing of coal at refining plants has grown almost two times. Along with the exploitation of old coal mining centers (Pechora, Kuznetsk, Kansk-Achinsk, Gorlovka, and Minusinsky coal basins), new deposits are being developed in Eastern Siberia and the Far East (Ulug-Khem basin and Elga coal deposit in the South Yakutsk coal basin) and the Arctic zone of Russia (Upper Alkatvaamsk and Amaamsk coal deposits in the Chukotka Autonomous District, Malolembersk, Nizhnelemberovsk, and Syradasay deposit in the Taimyr coal basin). The share of exports in total coal supplies increased from 33.4% to 54.0%. The primary destinations are Japan, South Korea, China, India, Malaysia, and Vietnam. However, the increase in the volume of coal supplies is hampered by the low capacity of the railway network in Siberia and the Far East (almost no railways in the Arctic), notably the Trans-Siberian Railway for supplies to China. Owing to the underdevelopment of the logistics system, coal producers suffer delays and higher operating costs for transportation and transshipment of coal at railway terminals and seaports. The decline in demand for coal both in the domestic and international markets amid growing competition between coal and more environmentally friendly natural gas poses a significant threat to the development of the conventional coal industry.

3.2.4 Electricity

Electricity generation has increased by 5.3% compared to 2008, consumption by 5.4%, and installed capacity of power plants by 11.0%. The development of the electric power industry is supported by the growth in demand for electricity from the transport sector, housing and utility services, and the expansion of energy-intensive industrial production in the Far East (cross-border collaboration with China and other countries in Asia). The growth of the electric power industry is hindered by the poor automation of technological processes and the increase in the vulnerability of power-generating facilities amid the complication of systems and algorithms for managing these facilities.

3.2.5 Nuclear energy

Electricity generation from nuclear power plants has increased by one-fourth since 2008. Russia's first floating nuclear thermal power plant (70 MW) was installed in Pevek, Chukotka Autonomous District to replace the retiring capacities of the Bilibino nuclear power plant in the Chaun-Bilibino energy hub. Russia leads the development of a new nuclear energy technology based on the parallel operation of thermal and fast neutron reactors within a single closed nuclear fuel cycle. Such a system contributes to solving the problems of nuclear fuel reproduction, minimizing radioactive waste, and nonproliferation of nuclear materials. The main issues and risks of the nuclear sector are associated with high safety costs and the need to manage nuclear fuel and radioactive waste (taking into account the environmental safety requirements and regulations). Furthermore, the share of cost-effective uranium reserves in the total mineral deposits in Russia accounts for only 7%.

3.2.6 Renewable energy

Electricity generation from hydroelectric power plants has grown by 15.8%. Hydropower plants, including pumped storage power plants, account for about one-fifth of the country's total power-generating capacities. Russia's huge hydropower potential (about 9% of the world's total) conditions the development of the renewable energy sector. The installed capacity of small hydropower plants, solar power plants, and wind farms integrated into the Unified Energy System of Russia is 1.2, 0.834, and 0.184 GW, respectively. The State Program on Energy Efficiency and Development of the Energy Sector (Government of the Russian Federation, 2014) states that Russia will have installed 2.24 GW of new solar energy facilities, 3.42 GW of wind farms, and 0.21 GW of small hydropower power plants by 2024 (Fig. 3.1). By 2035 the share of renewable sources in Russia's total energy mix will reach 6% (4.5% in

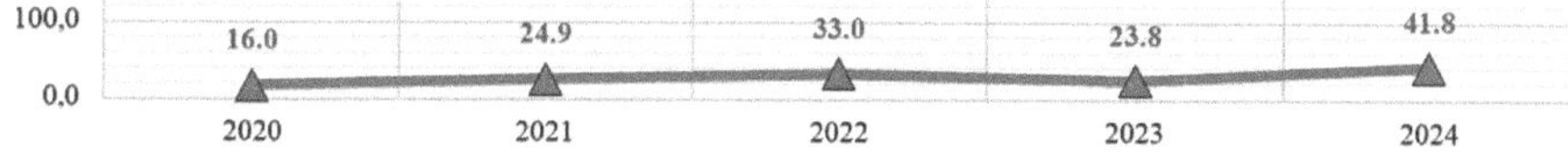

FIGURE 3.1 Introduction of new renewable energy capacity in Russia until 2024 (in megawatts) (Government of the Russian Federation, 2009). Source: *Authors' development based on Government of the Russian Federation. (2009). Decree #1-p from January 8, 2009, (updated on June 1, 2021).* On the approval of the main directions of the state policy in the field of improving the energy efficiency of the energy sector based on the use of renewable energy for the period until 2035. *<http://government.ru/docs/all/66930/>; Accessed 07.12.21.*

2024). The critical challenges to developing the hydropower sector in Russia include the lack of regulation of the legal status of reservoirs for hydropower purposes and the growing safety costs of hydraulic facilities. The construction of hydropower plants takes a long time, especially in the High North. One of the barriers to scaling up renewable energy sources is their weaker competitiveness and lower economic feasibility in contrast with abundant hydrocarbons.

3.3 Government regulation framework

3.3.1 Overall regulation

The Energy Strategy of the Russian Federation for the Period Until 2035 (Government of the Russian Federation, 2020b) outlines contemporary approaches to developing the fuel and energy sector in Russia:

- Establishing energy security of the country and its administrative entities, in particular in geostrategic territories
- Priority meeting of domestic demand for energy and energy-related goods
- Transitioning to clean and resource-saving energy
- Encouraging competition in the domestic fuel and energy market
- Rational use of natural resources and energy efficiency
- Maximum possible use of equipment produced in Russia
- Improving the efficiency and performance at all levels of management
- Maximizing the benefits of centralized power supply systems

According to the Energy Strategy (Government of the Russian Federation, 2020b), the primary measures to develop the fuel and energy complex in Russia include the following:

- Efficient support of the national economic development needs with appropriate production and export of energy and fuel. This means satisfaction of domestic demand for all types of energy and energy-related products and services, including within the framework of the implementation of national programs and projects. On one hand, the degree of satisfaction depends on the accessibility and usefulness of energy products and services to consumers. On the other hand, it also depends on the economy's energy efficiency.
- Spatial and regional development of the energy sector. That includes transforming and optimizing the energy-generating facilities and power grid infrastructure (taking into account the evolution of domestic and world markets, political and economic integration, and changes in international relations and regulations).

- Technological self-sufficiency and international competitiveness of the fuel and energy sector, that is, the development of competencies, localization of innovations, and production of technological complexes, equipment, materials, software, and relevant services in Russia.

Key measures have been identified in the main sectors of the fuel and energy complex to ensure stable production of oil, natural gas, and coal and the generation of energy from renewable sources. In the oil sector, development measures include more intensive exploitation of mature fields, exploration and development of small fields, stripped and highly waterlogged wells, and hard-to-recover reserves (including the Bazhenov suite) as well as the setting up of polygons for developing and testing technologies for extraction of hydrocarbons from hard-to-recover deposits (Fig. 3.2). Russia aims to modernize oil refineries (establishing over 50 secondary oil-refining units) to overtake most industrialized oil-producing countries. The goal is to increase operational efficiency and ensure the economic feasibility of oil refineries through automation, digitalization, and improved performance.

The future of the natural gas industry essentially pivots on the LNG sector. Within a few years, Russia aims to grow into one of the world's top producers and suppliers of LNG. There are plans to develop production and stimulate consumption of natural gas vehicle fuel (including fuel made of LNG). To achieve these goals, Russian gas tycoons such as Gazprom and Novatek must increase output up to 795–820 billion m^3 by 2024 and then further to 860–1000 billion m^3 by 2035. State support measures in the natural gas sector include the modernization and construction of supporting infrastructure (seaports, transport infrastructure, power grid lines, etc.), the establishment of the LNG cluster in the Yamal and Gydan peninsulas, and the launch of specialized hubs for transshipment, storage, and supply of LNG in the Arctic (in Murmansk and Kamchatka, the western and eastern ends of the Northern Sea Route (NSR), respectively). Russia also prioritizes the development of small-scale LNG facilities across the Arctic to improve the energy security of remote territories and industrial sites.

The coal industry requires a radical transformation in view of the global transition to low-carbon energy and climate change mitigation efforts. Since the early 2000s, Russia has been phasing out inefficient coal extraction and processing facilities, introducing innovative technologies (automation and robotization in mining, processing, and transportation), and modernizing equipment and technologies for storage and transshipment of coal taking into account international environmental regulations and decarbonization targets.

Development measures in the sphere of electric power include the optimization of the structure of generating capacities as part of the General Scheme for Allocation of Electric Power Facilities while maintaining the combined generating of electricity and thermal energy. The

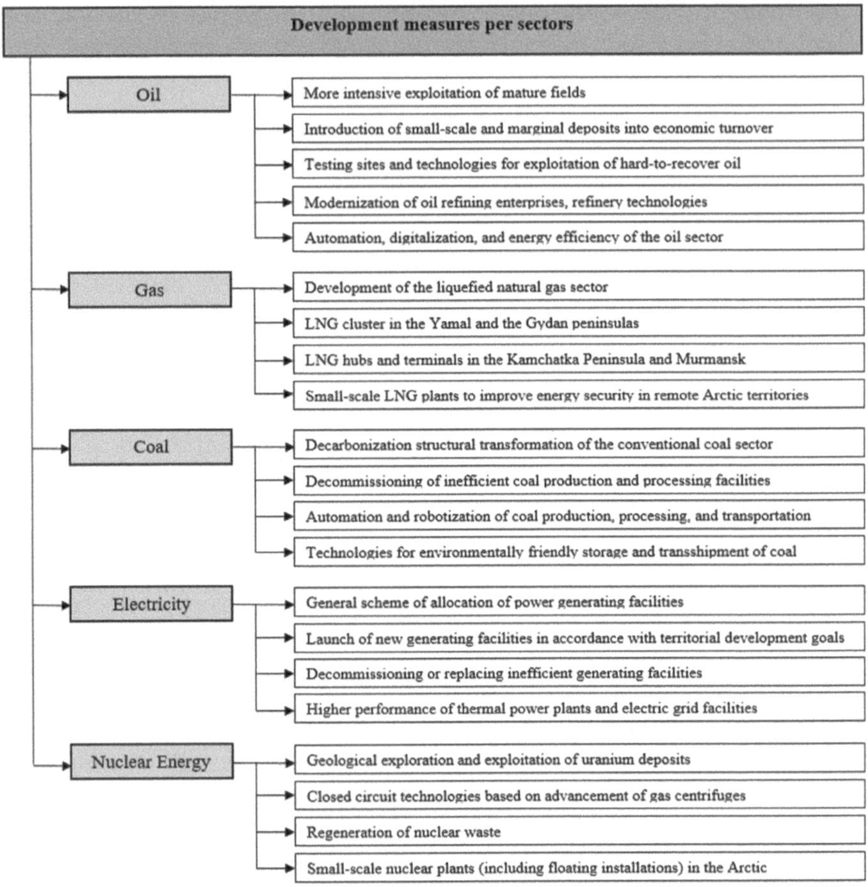

FIGURE 3.2 Measures to ensure steady growth of the energy sector in Russia. Source: *Authors' development based on Government of the Russian Federation. (2020b). Decree #1523-r from June 9, 2020.,* Energy strategy of the Russian Federation for the period until 2035. *<from https://minenergo.gov.ru/node/1026>; Accessed 08.12.21.*

launch of new generating facilities is synchronized with the growing demand for electricity in emerging industrial clusters and in remote territories across the Arctic zone. The government also aims to reduce excess capacity in the Unified Energy System and maintain an average level of power reserves by decommissioning or replacing inefficient generating facilities. In general, the regulation aims to improve the technical and economic performance of thermal power plants and the power grid system.

The growth of the nuclear energy sector requires an adequate supply of raw materials. The latter, in turn, involves geological exploration and development of uranium deposits in Russia. It is also necessary to develop closed-circuit technologies based on the new generation of gas centrifuges, increase the economic efficiency of nuclear fuel production and its components, and create industries to produce new types of nuclear fuel. Russia plans to build closed-circuit enterprises to treat and regenerate spent nuclear fuel and radioactive waste and make fuel from regenerated nuclear materials. The construction of low-power nuclear power plants to supply remote and isolated Arctic territories is outlined among the Energy Strategy's development priorities (Government of the Russian Federation, 2020b).

To spur innovation-based growth of the renewable energy sector, the government aims to incentivize competition; develop infrastructure for generation, accumulation, and transmission of energy; and improve the quality and efficiency of governance (Fig. 3.3). The competition sector includes regulation of prices for renewable energy in the wholesale market; government procurement of a certain amount of renewable energy; improvement of the

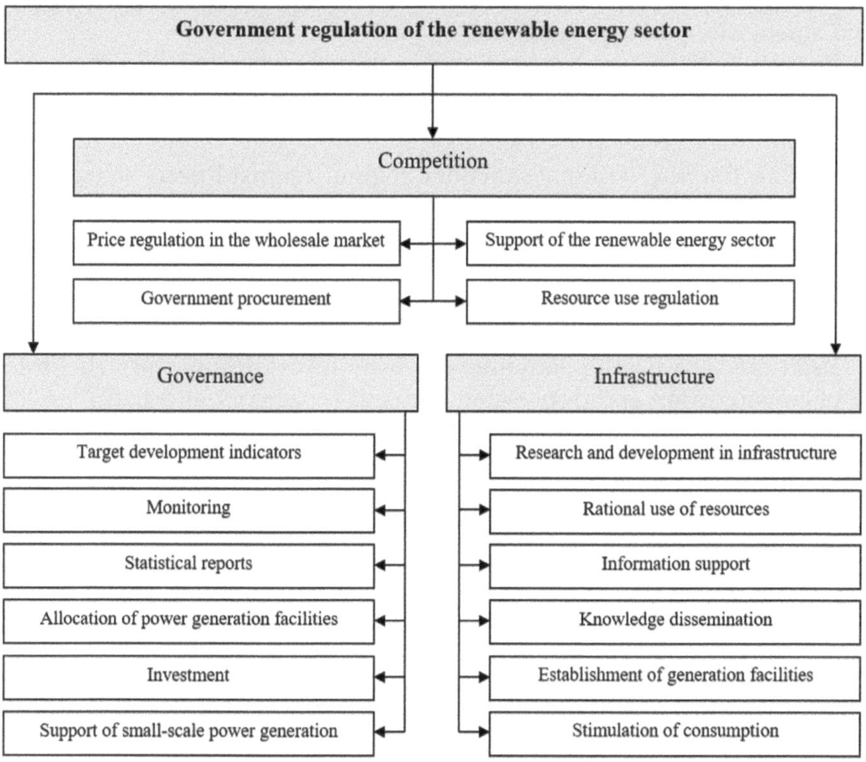

FIGURE 3.3 Government regulation of the renewable energy sector in Russia until 2035. Source: *Authors' development based on Government of the Russian Federation. (2009). Decree #1-p from January 8, 2009, (updated on June 1, 2021).* On the approval of the main directions of the state policy in the field of improving the energy efficiency of the energy sector based on the use of renewable energy for the period until 2035. *<http://government.ru/docs/all/66930/>; Accessed 07.12.21.*

legal regime for the use of natural resources; and state support for the renewable energy sector from federal and local budgets. The government prioritizes the development of infrastructure for generation and transmission of hydropower, solar, wind, and other kinds of renewable energy by developing scientific and technological services for the development of renewable energy, rational use of the potential of the domestic renewable energy sector, information support and introduction of advanced information technologies, training programs and dissemination of knowledge, documentation for the construction and operation of power generation facilities, and stimulation of consumption of renewable energy and goods produced thereof. In the sphere of governance, the strategy outlines the improvement of the system of target indicators and reports, monitoring the achievement of targets in the renewable energy sector, continuing updating of the allocation scheme of renewable energy facilities, attraction of investment in the establishment of power generation facilities, and support of small and medium enterprises and small-scale power generation.

Russia's fuel and energy complex is deeply integrated into the global market. Hence it faces considerable global environmental, economic, social, and political challenges. Some of the threats, however, are specific to Russia:

- In 2022, the international sanctions regime against Russia sharply tightened due to the Russia-Ukraine conflict. It directly affects not only the cooperation on energy projects (including the flagship Yamal LNG and the about-to-launch Arctic LNG 2 projects), but also the export of energy resources from Russia in general.
- Slowdown in economic growth since the early 2010s (a sharp decline amid the COVID-19 pandemic) has been transforming the structure of consumption and depressing demand for energy and fuel. The overproduction of hydrocarbons pushes prices down.
- Low demand for energy and fuel in the domestic market disincentivizes innovations and decreases growth prospects. Russia increasingly depends on exporting conventional energy and hydrocarbons.
- Critical dependence of Russian energy companies on imports of technologies, equipment, materials, services, and software in emerging sectors of the energy market.
- Lack of investment resources resulting from the government control of tariffs in the energy sector, international sanctions on long-term foreign investment and credit, and underdeveloped venture lending mechanisms.
- Global challenges to the scientific and technological development of the energy sector, such as the evolution of global and local energy systems, energy efficiency of economic growth, increase in generation and accumulation of energy, its transmission and efficient environmentally clean use.

The spatial development of any big country is challenged by imbalances in the allocation of energy production and energy consumption centers. In Russia, the mismatches are especially significant. An unprecedentedly large and constantly growing volume of expensive transportation of energy and fuel over long distances increases transaction costs. Population and industrial production are concentrated in the European part of the country (over 60% of the total energy consumption). In contrast, energy production has shifted northward and eastward. Siberia, the Far East, and the Arctic zone of Russia now account for about 80% of the country's total energy generation.

3.3.2 Arctic zone of Russia

The implementation of the spatial development priorities of Russia's energy policy involves a significant increase in the production and processing of all types of energy resources in the Russian Arctic, Eastern Siberia, and the Far East. The aim is to increase the production of transportable energy-intensive high-processed goods and to improve transport, economic, and social infrastructure of the energy sector in eastern and northern parts of the country. Development of the hydrocarbon resource potential of the continental shelf of the Arctic Ocean and northern territories is a pivotal geopolitical and technological challenge for Russian oil and gas companies. Even today, the Russian Arctic accounts for 80% of natural gas output and 17% of oil output (including gas). The Russian section of the continental shelf contains more than 85.1 trillion m^3 of natural gas and 17.3 billion tons of oil (including gas condensate). The progressing climate change is easing access to these resources and opening up new transportation routes in the Arctic Ocean and across Eurasia.

The Arctic zone of Russia comprises nine territories (Fig. 3.4). Four administrative entities (Murmansk Oblast, Nenets Autonomous District, Yamal-Nenets Autonomous District, and Chukotka Autonomous District) are located in the Arctic. In the remaining five entities, individual districts are recognized as belonging to the Arctic zone (in Arkhangelsk Oblast: the districts of Mezensky, Novaya Zemlya, Onezhsky, and Primorsky and the cities of Arkhangelsk, Novodvinsk, and Severodvinsk; in Komi Republic: Vorkuta city; in Krasnoyarsk Krai: the districts of Taimyrsky Dolgano-Nenetsky and Turukhansky and Norilsk city; in Republic of Karelia: the districts of Belomorsk, Loukhsky, and Kemsky; in Republic of Sakha (Yakutia): the districts of Abyisky, Allaikhovsky, Anabarsky, Bulunsky, Verkhnekolymsky, Verkhoyansky, Zhigansky, Momsky, Nezhnekolmsky, Oleneksky Evenki, Srednekolymsky, Ust-Yansky, and Eveno-Bytantaysky).

Among major energy-related threats to the spatial development of the Arctic, the Strategy for the Development of the Arctic Zone of the

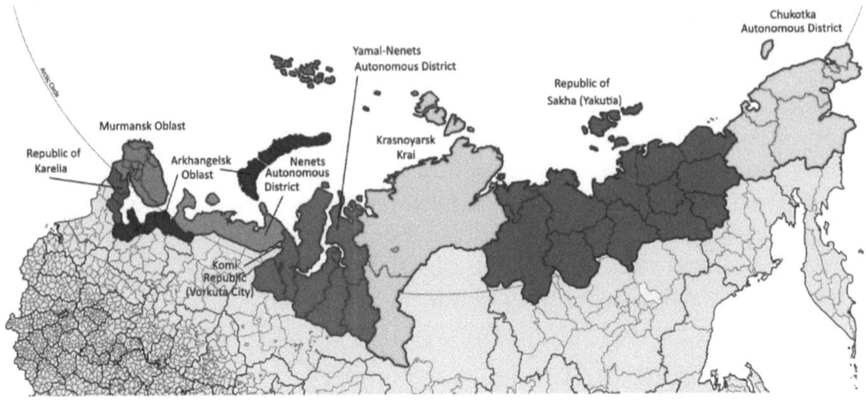

FIGURE 3.4 Land territories of Russia's Arctic zone. Source: *Authors' development.*

Russian Federation and Ensurance of National Security Until 2035 (President of the Russian Federation, 2020b) emphasizes a high share of local electricity generation based on the use of inefficient (in economic terms) and dirty (in environmental terms) diesel fuel. In light of sustainable development efforts, the problem is exacerbated by the low level of investment in fixed assets for the protection and rational use of natural resources against the background of intensive climate warming (two times faster in the Arctic than in the rest of the planet). The achievement of sustainable development goals by improving the efficiency of the energy sector and the rational use of energy resources assumes the following:

- A special economic regime in the Arctic to incentivize the transition to a circular economy, spur private investment in geological exploration, create new industrial facilities and modernize existing ones, develop new oil and gas provinces and fields of hard-to-recover hydrocarbon reserves, and increase the volume of deep oil refining and production of LNG and gas chemical products
- Providing investors with state support in their capital investments in energy and engineering infrastructure facilities, including the infrastructure of gas and water supply, pipeline transport, and communication systems required for the implementation of new investment projects
- Development and implementation of a geological survey program in the Arctic
- Elaboration and development of a new model for the performance of resource projects on the continental shelf, providing for the expansion of private investors' participation in such projects while maintaining state control

- State support aimed at the creation and development of innovations in the oil and gas sectors, including LNG technologies and the exploration of the continental shelf
- Use of LNG in sea and river transport in the NSR and energy supply to industries and residential areas
- Development and implementation of engineering and technical solutions to ensure stable operation of industrial, transport, and residential infrastructure amid increasingly unpredictable climatic events
- State support aimed at improving the efficiency of electricity generation in remote and hard-to-reach territories, including using small-scale LNG facilities, renewable energy sources, and local fuels
- Providing indigenous communities with mobile sources of energy supply and means of communication in traditional places of their residence and economic activity

The strategy is carried out in three stages (Table 3.2). We defined five activities in the sphere of energy production and supply.

TABLE 3.2 Implementation of the strategy by stages until 2035 (energy sector).

Activities	2020–24	2025–30	2031–35
Integrated development of mineral resource centers and adjacent territories	Pilot projects	Program implementation	Modernization of the urban environment and social infrastructure
Replacement of inefficient diesel power generation in isolated and hard-to-reach territories with generation based on LNG, renewable energy sources, and local fuel	Program implementation		Program completion
Exploration and exploitation of energy resources on the continental shelf (Arctic Ocean)	Application of a new model for the implementation of economic projects on the continental shelf		Capacity building of enterprises
Innovations in the energy sector		Putting new equipment into operation and production, including portable energy sources	
Elimination of harmful effects of large-scale energy projects in the Arctic	Rehabilitation of territories where nuclear fuel and radioactive waste are located		

Authors' development based on President of the Russian Federation. (2020a). Decree #164 from March 5, 2020. On the foundations of the state policy of the Russian Federation in the Arctic till 2035. *<http://www.kremlin.ru/acts/news/62947>; Accessed 07.12.21. President of the Russian Federation. (2020b). Decree #645 from October 26, 2020.* On the strategy for the development of the Arctic zone of the Russian Federation and ensurance of national security till 2035. *<https://docs.cntd.ru/document/566091182>; Accessed 07.12.21.*

In individual territories that comprise the Arctic zone of Russia, the above five types of activities are being implemented differently, owing to a number of factors and inherent characteristics of the territories, such as climate, geographical location, availability of natural resources, exploitation of those resources, transport and infrastructure interconnectedness of territories with each other and with the rest of the country, development of the local energy sector, spatial development parameters, and many more (Gao & Erokhin, 2021). A review of regional development strategies of nine territories allowed us to identify individual development priorities in four key sectors: coal, oil, natural gas, and renewable energy (Table 3.3). The exploitation of energy resources on the continental shelf should be considered separately from conventional oil and gas sectors, since such projects are of national and not only regional importance. Currently, Russia operates only one Prirazlomnaya oil platform (Gazpromneft), but geological exploration that has been conducted on the shelf aims to develop new oil and gas offshore fields to increase oil and LNG output. The infrastructure for the generation, accumulation, and transmission of energy between geographically separated industrial sites and settlements should also be considered separately from the other five sectors.

The vision for developing the energy complex is obviously based on the natural and acquired competitive advantages that a particular territory possesses. Thus Republic of Karelia intends to build cascades of small hydropower plants along numerous rivers and lakes (Government of the Republic of Karelia, 2018; Government of the Russian Federation,

TABLE 3.3 Development priorities in the energy sector.

Territories	Coal	Oil	Natural gas	Offshore energy projects	Renewable energy	Power generation and transmission
Arkhangelsk Oblast						
Chukotka Autonomous District						
Komi Republic						
Krasnoyarsk Krai						
Murmansk Oblast						
Nenets Autonomous District						
Republic of Karelia						
Republic of Sakha (Yakutia)						
Yamal-Nenets Autonomous District						

Authors' development based on President of the Russian Federation (2020a, 2020b); Governor of Murmansk Oblast (2014); Government of Murmansk Oblast (2013); Assembly of Deputies of Nenets Autonomous District (2019); Government of Chukotka Autonomous District (2014); Legislative Assembly of Yamal-Nenets Autonomous District (2011); Governor of Yamal-Nenets Autonomous District (2018); Government of the Republic of Karelia (2018); Government of the Russian Federation (2020a, 2020b, 2021); Assembly of Deputies of Arkhangelsk Oblast (2019); Government of the Komi Republic (2019); Government of Krasnoyarsk Krai (2020); Head of the Republic of Sakha Yakutia (2020).

2020a), while the Strategy of Social and Economic Development of Arkhangelsk Oblast Until 2035 (Assembly of Deputies of Arkhangelsk Oblast, 2019) envisages the establishment of a full-cycle timber-processing industry and the advancement of technologies for the production of biofuels from waste generated by timber processing. Based on the significant technical and technological potential of its shipbuilding enterprises, Arkhangelsk Oblast is developing the construction of large-scale facilities and the production of equipment for oil and gas projects on the Arctic Ocean shelf. Russia's Northwest Arctic is one of the country's centers of Arctic shipbuilding and marine engineering (Gao & Erokhin, 2020a). Parallel to Arkhangelsk Oblast, the neighboring Murmansk Oblast establishes a center for constructing large-capacity offshore structures for the production, storage, and shipment of LNG. The local government also supports enterprises engaged in the repair and maintenance of marine machinery and equipment used to develop offshore deposits (Governor of Murmansk Oblast, 2014). In addition, Murmansk Oblast prioritizes geological surveys of mineral resources in the Kola Peninsula along with the development of existing clusters for the extraction and refinery of minerals and the establishment of new ones. The Strategy of Social and Economic Development of Murmansk Oblast Until 2020 and for the Period Until 2025 (Government of Murmansk Oblast, 2013) envisages the development of power-generating infrastructure, including the replacement of outdated fuel combustion facilities with more environmentally friendly solutions that use other types of energy resources to generate electricity or produce heat.

The central part of the Russian Arctic (Yamal-Nenets Autonomous District, Krasnoyarsk Krai, and Komi Republic) is the most advanced part of the Arctic zone in terms of developing natural resources. The Strategy of Social and Economic Development of Yamal-Nenets Autonomous District Until 2020 (Legislative Assembly of Yamal-Nenets Autonomous District, 2011) emphasized the importance of exploring natural gas fields in the Ob basin and the Ob Bay along with the development of pipeline infrastructure. Strategic projects are the Novoportovsk oil and gas condensate mineral resource center, the Bovanenkovo gas condensate mineral resource center, and the Tambey group of fields as well as the development of offshore fields. Other priorities include the maintenance and development of gas and oil pipeline networks and the development of gas and oil mineral resource centers connected to pipelines in the Nadym-Pur and Pur-Taz oil and gas regions, including the use of new technologies for the extraction and development of underlying layers and hard-to-extract oil reserves (Gao & Erokhin, 2021). The rapid growth of the LNG industry in the Russian Arctic poses new challenges, including the expansion of LNG production on the Yamal and Gydan peninsulas and the development of technologies for using low-pressure natural gas in industrial production, including gas compression technologies (Governor of Yamal-Nenets Autonomous District, 2018). Regarding infrastructure development, the focus is on the expansion of the central power supply zone by connecting settlements to a single energy system as well as the development of oil and gas services by creating industrial zones in the largest settlements.

The conventional energy sector is a pivotal factor of economic growth in most of the nine Arctic territories in Russia. Nenets Autonomous District establishes gas condensate mineral resource centers, such as the Korovinsky and Kumzhinsky gas condensate fields and the Vaneyvissky and Layavozhsky oil and gas condensate fields. The Strategy of Social and Economic Development of Nenets Autonomous District Until 2030 (Assembly of Deputies of Nenets Autonomous District, 2019) also prioritizes the development of the Varandeysky, Kolguyevsky, Kharyago-Usinsky, and Khasyreysky oil and mineral resource centers. Krasnoyarsk Krai has set up new production facilities at the Zapolyarnaya mine and is modernizing the existing ones. On the basis of coal fields in the West Taimyr, the coal industry cluster has been established to export coal via the NSR. The Strategy of Social and Economic Development of Northern and Arctic Territories and Support of Indigenous People in Krasnoyarsk Krai Until 2030 (Government of Krasnoyarsk Krai, 2020) also emphasizes the establishment of the West Taimyr coal industry cluster to export coal via the NSR. The Republic of Karelia aims to stimulate deep processing of coal raw materials and coal chemistry in coal mineral resource centers in the Pechora coal basin (Government of the Komi Republic, 2019). It also intends to develop oil and gas mineral resource centers in the Timan-Pechora oil and gas province, including setting up oil- and gas-processing facilities.

Territories in the Eastern sector of the Arctic zone (Republic of Sakha (Yakutia) and Chukotka Autonomous District) struggle with extremely weak territorial interconnectedness and a low level of energy infrastructure development. Therefore spatial development by forming networks of interconnected industrial and energy clusters is a key sustainable development goal in this part of the Russian Arctic. The Strategy of Social and Economic Development of the Arctic Zone of the Republic of Sakha (Yakutia) Until 2035 (Head of the Republic of Sakha Yakutia, 2020) prioritizes (1) integrated development of the Anabar and Lena basins, development of mineral resource centers, such as the Taimylyr coal deposit and the West Anabar oil and mineral resource center; (2) development of territories in the Indigirka River basin and ensuring their energy security by means of the exploration of the Krasnorechensk coal deposit; (3) development of territories in the Kolyma River basin through the exploitation of the Zyryansk coal mineral resource center.

In Chukotka, the Anadyrsky, Egvekinotsky, and Chaun-Bilibinsky power hubs operate separately. The Strategy of Social and Economic Development of Chukotka Autonomous District Until 2030 (Government of Chukotka Autonomous District, 2014) prioritizes the modernization of the Chaun-Bilibino power generation system and the development of the Bering coal mineral resource center.

3.4 Energy and innovations

Common challenges for stable and sustainable development of the energy sector in the Arctic include extreme climate, fragile ecosystems, and poor accessibility of many deposits due to low temperature, permafrost, ice, and so on. Many scholars (Balakireva et al., 2018; Gerstlberger, Knudsen, Dachs, & Schroter, 2016; Irandoust, 2019; Melnik, Naoumova, Ermolaev, & Katrichis, 2021; Miao, Fang, Sun, Luo, & Yu, 2018; Song & Zhang, 2020; Urpelainen, 2021) agree that the introduction of innovations in all segments of the fuel and energy complex (not only renewable, but also conventional oil and gas industries and nuclear and electric power) not only increases the efficiency of energy projects, but also reduces the environmental risks that are associated with them. The Energy Strategy (Government of the Russian Federation, 2020b) postulates that Russia needs accelerated transition (a modernization breakthrough) to a more efficient, flexible, and sustainable energy sector that is able to respond adequately to existing challenges and threats. Such a modernization shift should be based on the following four types of qualitative transformations:

- Structural diversification, in which carbon energy is supplemented by noncarbon sources, centralized energy supply is supplemented by decentralized mobile sources, and export of energy resources is supplemented by the export of technologies, equipment, and services
- Digital transformation and intellectualization of the fuel and energy complex
- Optimization of spatial allocation of the energy-generating infrastructure, including the establishment of oil and gas mineral resource centers and petrochemical complexes in Eastern Siberia, the Far East, and the Arctic zone along with the development of infrastructure for transportation and transmission of energy
- Reduction of the negative impact of the fuel and energy complex on the environment and adaptation of the energy-related technologies to pursuing climate change mitigation and decarbonization goals

According to the Energy Strategy (Government of the Russian Federation, 2020b), the contribution of the energy sector to the achievement of national goals and the solution of strategic tasks consists in accelerating technological development, digitalizing the energy sector, and increasing Russia's export potential:

1. Acceleration of technological development, increasing the number of enterprises that are engaged in technological innovations in the energy sector up to 50% of their total amount:
   a. Coordination of state programs of scientific and technological development, programs of innovative development of public companies, and publicly funded fundamental and applied research projects
   b. Establishment of competence centers in priority areas of technological development of the fuel and energy complex
   c. Establishment of engineering centers and testing sites for the introduction of innovative technologies and modern materials, samples of new equipment, most advanced technologies, and new products
   d. Development of venture business in the sphere of innovation and support for commercialization of research and development in the energy sector
   e. Improvement of state support mechanisms for innovative projects, including those in the field of cross-cutting digital technologies and platform solutions
2. Accelerated introduction of digital technologies:
   a. Development of a system for management, coordination, and monitoring of digital transformation in the fuel and energy complex
   b. Introduction of digital technologies in public administration and control and supervisory activities
   c. Implementation of pilot projects to introduce digital technologies and industry platform solutions
3. Creation of a high-performance export-oriented sector based on advanced technologies and highly qualified personnel:
   a. Increase in export of energy resources by 9%–15%
   b. Increase in LNG output by 240%–340%
   c. Establishment of six petrochemical clusters
   d. Production and consumption of hydrogen
   e. Russia to rank among the world's top producers and suppliers of hydrogen
   f. Production, long-term storage, and supply of helium
   g. Infrastructure for transportation of liquid helium in the Far East

The development priorities in the field of the innovative potential of the energy sector in the Arctic include establishing an innovative research center for conducting targeted R&D and advancing international scientific and technical cooperation. It is necessary to conduct logistics analysis in order to identify rational ways and means of multimodal transportation of hydrocarbons. Innovative marine engineering technologies are needed to scale up the fleet of icebreakers, ice-strengthened vessels, and LNG and oil tankers in the NSR (Gao & Erokhin, 2020a). Concurrently, it is necessary to rely on the sustainable development agenda and the territory-specific environmental safety priorities (further addressed later in the chapter). Both the government and commercial companies should focus on automating difficult industrial tasks by introducing innovative technologies. Examples are automatic diagnostics of the integrity of pipeline systems and automation of operation of compressor stations employed in the Nord Stream pipeline network. Ensuring continuous and comprehensive monitoring of safety and performance of power generating and transmitting facilities is particularly important. For example, Gazprom Space Systems operates a satellite system to monitor Gazprom pipeline networks and other facilities. One of the factors of stable development of circumpolar territories is a stable and reliable supply of electricity, heat, fuels, and other types of power. In remote territories, power supply stability can be achieved by using small-scale mobile energy-generating solutions, such as renewable energy installations or low-power nuclear power plants. They can supply individual settlements or industrial sites (including offshore) with no need to establish extensive and expensive infrastructure (some of the solutions are discussed later in the chapter in the section on small-scale power generation). Finally, it is essential to ensure the safety of the transportation of hydrocarbons, fuel, and energy in the harsh climatic conditions and fragile ecosystems of the Arctic. Examples are technologies for the production of high-strength pipes (anticorrosion coating, concreting, etc.) or construction materials for the Arctic (resistant to thawing permafrost, hummocks, icing, extreme temperatures).

3.5 Energy and sustainable development

Despite the common recognition of the crucial importance of sustainable development efforts in the Arctic, many aspects of such development in the fuel and energy complex have remained poorly implemented. Numerous human-made disasters and technological emergencies in the Russian Arctic over recent years (oil and fuel spills, fires, and soil degradation) have caused severe environmental consequences (Morgunova & Kovalenko, 2021). The common causes of disturbance include low diversification of the Arctic economy (too much stress on certain types of resources in certain territories), lack of access to advanced technologies (energy-saving, low-carbon, environmentally friendly), accumulation of industrial and municipal waste, high cost of economic activity and life support in extreme climate conditions, and weak control over anthropogenic activities in the Arctic by environmental authorities (Gao & Erokhin, 2020b).

Seemingly, major sustainability-related problems in the Arctic include the reduction of biological diversity of aquatic and terrestrial ecosystems, pollution of air and inland and water areas with waste and harmful substances, violation of the integrity of Arctic ecosystems due to soil degradation and inappropriate land use, forced changes in the traditional way of using natural resources by indigenous peoples, and negative consequences of progressing climate change (Gao et al., 2021). At first glance, these problems are not directly related to the energy sector, so they cannot be addressed by any measures or innovative solutions related to energy. Still, the indirect impact of both anthropogenic activities in general and the energy sector in particular in generating various types of power for industrial and residential purposes is exceptionally significant

For example, adverse effects of climate change are inextricably linked to the ever-increasing consumption of energy in its various forms (Danubianu & Teodorescu, 2021; Jessoe & Moore, 2021; Kung & McCarl, 2018; Streimikiene, Balezentis, & Krisciukaitiene, 2012; Taseska, Markovska, & Callaway, 2012; Toke & Vezirgiannidou, 2013). More than two-thirds of territories that compose the Russian Arctic are located in the permafrost zone. Since the 1990s the areas of glaciation and snow cover have decreased by 16% and 12%, respectively. In some territories, coastal erosion is spreading by 10 m per year. The Arctic is particularly vulnerable to the climate change that is being seen today and to the more disturbing climate effects that are expected in the future. Some of these effects include an increase in winter temperature and evaporation, lower precipitation in summer, a change in the annual flow of rivers and its seasonal redistribution, a change in ice conditions in the Arctic Ocean and the mouths of northern rivers, and deeper seasonal thawing of permafrost. Regarding the energy sector, a less predictable climate aggravates risks of economic activities in coastal zones (erosion, flooding, waterlogging) and on the continental shelf (ice, storms, waves, winds) and threatens the stability of inland power transmission and the safety of shipping. Engineering, construction, and operational costs are rising significantly in all industries and types of economic activity across the Arctic zone.

Industrial development of circumpolar territories, exploitation of oil and gas fields in the Arctic, and transportation of mineral resources disturb wildlife habitats (mammals, fish, birds, reindeer, polar bear, polar fox, waterfowl). Irreversible changes occur to freshwater reservoirs and traditional fishing of indigenous peoples (shallowing of rivers, destruction of spawning grounds, drainage and eutrophication of lakes). Traditional reindeer pastures and seasonal migrations routes have been degrading. The most disturbed territories are those adjacent to major oil, gas, and coal industrial sites (Timan-Pechora, Novozemelsk, Yamal, and Sredne-Obsky oil and gas basins and coal mines in Vorkuta). The industrial development of the Arctic and climate change are mutually dependent and push each other. Warming improves access to new mineral deposits, the development of which, in turn, harms the environment and climate and disturbs fragile Arctic ecosystems. Melting of sea ice in the Arctic Ocean entails intensifying navigation and aggravates risks of accidents and pollution of water areas.

Climate change is one of the triggers for the decline in the biodiversity of aquatic and terrestrial Arctic ecosystems. A decrease in the area of permafrost and sea ice, a shorter ice period, thermal and coastal erosion, disbalance of the ocean-land atmospheric exchange, the rising sea level and river runoff, waterlogging, and drying up lakes all transform wildlife habitats, degrade ecosystems, and induce invasions (or artificial introduction) of alien species into the region. Arctic biodiversity is particularly endangered by intensive growth of the fuel and energy sector, including through the following:

- Pollution of terrestrial and marine habitats with petrochemicals, oil, refuse fuel, and drilling waste
- Expansion of marine and onshore oil and gas infrastructure and emergencies due to spills and leakages of oil, fuel, and petrochemicals during extraction, storage, and transportation
- Unregulated construction of the power-generating and transmitting facilities, geological exploration of mineral resources, and transportation and transmission of hydrocarbons, fuels, and other energy resources in various forms

The problem of pollution with waste from invasive anthropogenic activities has several components. These are disposals of liquid and solid radioactive waste, including spent nuclear fuel, the discharge of liquid and solid radioactive waste in the shallow bays of the Kara Sea and the Barents Sea, and on-site burial of waste due to high transportation costs. Annually, up to one billion tons of dumping rocks and solid waste are generated by industrial enterprises in the Arctic zone, primarily in Murmansk Oblast, the lower reaches of the Pechora River (Nenets Autonomous District), Yamal-Nenets Autonomous District, Norilsk industrial area, the north of the Republic of Sakha (Yakutia), and gold mining sites on the Chukotka Peninsula.

Soil degradation and land use disturbance disrupt the integrity of Arctic ecosystems. Transformed tundra lands account for only 1%–3% of the total area of the Russian Arctic mainland. However, the soil cover is dramatically destroyed around large industrial enterprises and deposits. Within a radius of tens of kilometers, natural landscapes and live vegetation have been transformed by people. Only Arctic deserts (except for coastal areas), where industrial development remains poor, have remained relatively undisturbed. According to numerous studies (Brekhuntsov, Petrov, & Prykova, 2020; Collins & Kumral, 2020; Huntington et al., 2007; Makarov & Stepanov, 2015; Medvedeva, 2015; Trump, Kadenic, & Linkov, 2018), environmental degradation is caused by the extraction of hydrocarbons and minerals but also by the construction of transport and industrial infrastructure, extreme climate events, permafrost thaw, deforestation, and forest and tundra fires. Disturbance foci emerge as a result of landscape degradation in industrial localities and then spread as a result of the construction of transport infrastructure and power grid lines and merge into large clusters of disturbed lands and water areas.

The environmental problems that we have described induce radical changes in the traditional way of nature management and economic activities of indigenous people across the Arctic zone. Large-scale reindeer herding suffers from increasingly frequent glazes and ice rains that cause famine and cattle fall. Traditional pastures and seasonal migrations routes have been transforming as a result of the industrial development of territories. Climate change and anthropogenic impact cause irreversible changes in freshwater reservoirs and traditional fishing. The availability of conventional biological resources (the basis of the traditional nature management of the indigenous people) is decreasing as a result of the deeper spread of anthropogenic pollution into the Arctic (up to 40% of the traditional nature management area of indigenous communities). The most disturbed habitats are the Kola, Timan-Pechora, Novozemelsk, Vorkuta, Yamal, Sredne-Obsk, Norilsk, Anabar, Yano-Mndigirsk, Valkumeysk, and Bilibino districts.

According to the Strategy for the Development of the Arctic Zone (President of the Russian Federation, 2020b), key tasks in the field of environmental protection and environmental safety are accomplished through the implementation of the following measures:

- Adaptation of the Arctic economy and infrastructure to climate change
- Development of the system of state environmental monitoring
- Identification, assessment, and elimination of cumulative environmental damage
- Reduction of harmful emissions into the atmosphere and discharges into the water from industrial enterprises and municipal facilities
- Prevention of negative environmental consequences of development and exploitation of hydrocarbons and mineral resources

- Development of a state system for the prevention and elimination of emergencies, such as spills of oil, fuel, and petrochemicals in inland industrial sites, terrestrial transport infrastructure, and along the NSR
- Regular assessment of the impact of nuclear power facilities on the environment, ecosystems, and people
- Rational use of associated petroleum gas to reduce gas flaring
- State support for waste management in the Arctic zone and improvement of the hazardous waste management system

The Energy Strategy (Government of the Russian Federation, 2020b) provides for several measures in the energy sector designed to contribute to environmental protection and combating climate change:

- Adherence to the best available technology principle in the energy sector
- Creation of a national system for monitoring and reporting greenhouse gas emissions, including from energy- and power-generating facilities
- Consideration of risks of changes in climatic and hydrometeorological conditions along with common financial and economic parameters in assessing the feasibility and implementation of industrial and infrastructure projects
- Participation in the development of international environmental legislation and harmonization of Russia's regulations with international agenda
- Reduction of pollution, recycling and management of accumulated industrial waste, land reclamation, and other technical and organizational measures to compensate for environmental damage caused by the energy sector
- Removal of infrastructural, technological, and other restrictions on the rational use of associated petroleum gas and minimization of gas flaring
- Stimulation of research and support for the development of promising technical solutions aimed at reducing the negative impact of the energy sector on the environment and minimizing environmental risks
- State regulation of greenhouse gas emissions and other anthropogenic impacts of the energy sector on climate (in line with Russia's international obligations)
- Implementation of the provisions of the Stockholm Convention on Persistent Organic Pollutants (United Nations, 2001), including treatment or disposal of equipment and waste that generate or contain persistent organic pollutants
- Ensuring openness and accessibility of environmental information; timely informing of stakeholders about accidents, their

environmental consequences, and elimination measures; and strengthening interaction with public environmental organizations and societies
- Development and use of environmentally friendly, low-carbon, and resource-saving technologies for the production, transportation, storage, and use of energy resources, including clean coal technologies
- Increasing the fuel utilization factor in generating electric and thermal energy and reducing the specific consumption of equivalent fuel for the supply of electrical power;
- Increase in the production of high-quality motor fuels (including gas engine fuels) with improved environmental characteristics that comply with international norms and standards
- Advanced training of employees who are responsible for industrial and environmental safety of energy-generating facilities and processes

Ensuring sustainable development of circumpolar territories through the energy sector should focus on elaborating national environmental standards in the field of innovation with due regard to international experience. Examples are catalogs of the best available technologies for the Arctic or a system of indicators for assessing the impact of innovative technologies on the environment. The reduction of the environmental pollution risk of the energy sector is crucial for most of the territories in the Russian Arctic, particularly Yamal-Nenets Autonomous District (LNG production), Komi Republic (Vorkuta coal cluster), and Krasnoyarsk Krai (Norilsk industrial cluster). For instance, leakage risks can be reduced by using corrosion protection, vibration and noise insulation, radiation protection, encapsulation and waterproofing adjusted to the use in Arctic conditions. Examples of innovations in the sphere of oil spill response include the localization of oil spills (boom defense), collection of oil from the ice (trawling and sweeping), and thermal oil removal (burning or use of dispersants, sorbents, and coagulants). The focus should be made on constructing new industrial complexes for storage, treatment, utilization, and neutralization of especially hazardous waste generated in the Arctic. Examples of innovative technologies for processing and disposal of waste in Arctic conditions include plasma processing of waste, vitrification, glazing, and neutralization of hazardous radioactive waste by synthetic materials. It is necessary to develop an institutional framework for effective interaction between government agencies, private companies, educational institutions, and volunteer organizations to increase emergency response efficiency. Innovative training and competence development programs should capture key personnel involved in cross-border coordination of emergency responses in the Arctic. Implementing the above measures can improve the environmental safety of economic activity in the Arctic while reducing environmental risks from the anthropogenic impact on ecosystems and indigenous communities.

3.6 Sustainability through innovation

3.6.1 Russia's approach to boosting innovations in the energy sector

The strategic task for Russia is to ensure the sustainable development of circumpolar territories through research and innovations in the fuel and energy complex and related industries. A system approach involves the improvement of Russia's scientific and technological potential and the development of advanced technologies in the energy sector; promotion of innovative activities in the fuel and energy complex; modernization and competitiveness enhancement of Russian companies based on technologies, equipment, and materials of domestic production; and mutually beneficial international cooperation in the energy sector (Fig. 3.5). The following measures are called for to increase the innovation potential of the fuel and energy complex in Russia:

- Creation of a national system of technological forecasting that ensures coordination of produced forecasts with development strategies and programs in the energy sector and related industries

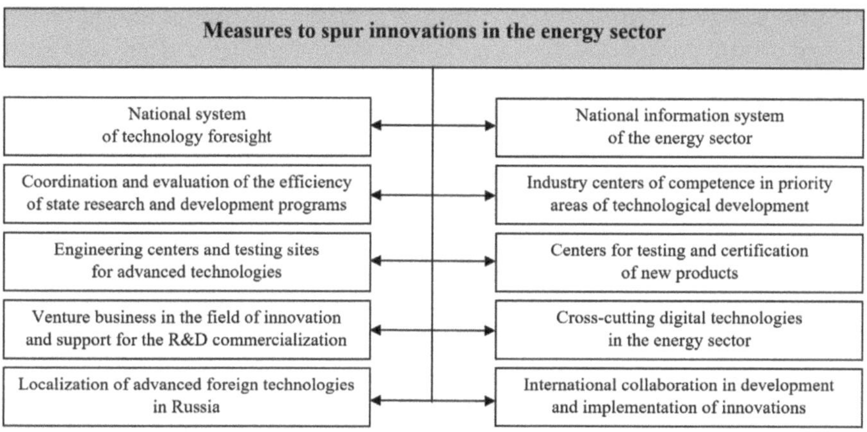

FIGURE 3.5 Measures to spur innovation-based development of the energy sector in Russia. Source: *Authors' development based on Government of the Russian Federation. (2020b). Decree #1523-r from June 9, 2020.* Energy strategy of the Russian Federation for the period until 2035. *<https://minenergo.gov.ru/node/1026>; Accessed 08.12.21.*

- Development of a state information system in the fuel and energy complex to produce high-quality statistical and analytical reports and forecasts
- Coordination and assessment of the performance of state programs in the fuel and energy complex and related industries, innovative development programs of individual public companies, and publicly funded fundamental and applied research projects
- Establishment of industrial centers of competence in priority areas of technological development in the energy sector
- Establishment of engineering centers and testing sites to condition the introduction of modern materials, samples of new equipment, innovative technologies, and new products
- Development of venture business in the field of innovation and support for the commercialization of the research and development results in the energy sector, including through continuous financing of promising projects at all stages of the innovation cycle
- Improvement of state support mechanisms for innovative projects, including projects in the field of cross-cutting digital technologies in the fuel and energy complex (including platform solutions)
- Localization of advanced foreign technologies required for stable operation and sustainable development of the fuel and energy complex, including through tax and customs tariff incentives
- International cooperation in the development and implementation of energy-related innovative technologies

The modernization of the energy sector relies on digitalization (or digital transformation, as stated in the Energy Strategy) (Government of the Russian Federation, 2020b). Russia's vision of digitalization in the energy sector comprises the development of digital services and solutions through management, coordination, and monitoring of digital transformation; the development and adjustment of legislation on the implementation of digital technologies in the energy sphere; the introduction of digital technologies in public administration, control, and supervision in the energy sector and related industries; and the selection and implementation of pilot projects for the introduction of digital technologies and platform solutions. Some of the measures are being implemented within the EnergyNet road map, a long-term National Technology Initiative program for the development of technology, business practices, standards, and communities to construct a new technological paradigm in the energy sector. Preferred developments include intelligent technologies and means of monitoring and diagnosing energy systems equipment; intelligent energy systems (digital substation, virtual power plant, intelligent electricity metering systems, highly sensitive equipment, power electronics, relay protection and automation devices, fast switching equipment); intelligent real-time management of energy consumption based on the integration of electrical and information networks (energy Internet); and cost-effective large-scale energy accumulation solutions. To establish reliable energy accumulation systems, Russia prioritizes the development of high-current accumulators with large cycling capacity and small-scale, low-cost drives; low-cost storage devices of high energy intensity, including those based on pneumatic or hydrogen systems; and high-performance water electrolyzers and small-scale systems for storage and transportation of hydrogen.

3.6.2 Renewable energy for the Arctic

In light of the climate-related effects of industrial exploration of the Arctic zone, reducing the carbon footprint turns out to be a core theme of the sustainability agenda (Aliev & Avramenko, 2017; Ellabban, Abu-Rub, & Blaabjerg, 2014). For now, renewable energy is one of the most efficient solutions to the emissions problem (Armeanu, Joldes, Gherghina, & Andrei, 2021; Neacsa, Panait, Muresan, & Voica, 2020; Popescu, Andrei, Nica, Mieilă, & Panait, 2019). The Arctic climate (extremely low temperatures, polar nights, storm winds, ice-covered waters, etc.) substantially complicates the expansion of renewable energy across the region (Berdin, Kokorin, Yulkin, & Yulkin, 2017). Still, many scholars (Bansal et al., 2005; Dincer, 2000; Gernaat et al., 2021; Muhammed & Tekbiyik-Ersoy, 2020; Xu, Assenova, & Erokhin, 2018) acknowledge that the use of renewable energy sources allows companies to reduce construction and maintenance costs in establishing energy-generating infrastructure or supplying power to remote industrial sites and settlements. Owing to the cold climate and highly dispersed population, building energy infrastructure from scratch is very expensive. In some territories it is even impossible. Maintaining and operating conventional energy facilities require uninterrupted service and continuous fuel and other supplies. The traditional energy system in the Arctic works on imported coal, oil, and diesel; that is, it emits a substantial amount of harmful substances into the air, water, and land. Renewable energy sources depend less on fuel delivery or do not depend on it at all. Therefore the carbon footprint of the renewable energy sector is negligible in comparison to the conventional burning of coal, oil, and even natural gas (Maiorov, 2020).

The Russian Arctic's most promising renewable energy sectors are hydropower, wind and solar energy, biofuels, and ocean energy. Wind speeds in coastal areas of the White and the Barents seas and Novaya Zemlya and Franz Josef Land archipelagoes reach 5–8 m/s. As the climate gets warmer, winds will become stronger and more frequent. Higher density of cold air compared to hot air preconditions higher energy efficiency of wind farms in the Arctic compared to those in temperate zones. Several wind power plants operate in the Arctic, such as the experimental four-turbine wind farm in Labytnangi in Yamal-Nenets Autonomous District; the ten-turbine wind power plant in Anadyr, Chukotka Autonomous District; and the three-turbine wind park in Tiksi, Republic of Sakha (Yakutia). In 2019, Enel Russia started construction of Russia's biggest wind farm above the Arctic Circle on the Kola peninsula, Murmansk Oblast (201 MW, 57 turbines). The Kola wind farm will generate about 750 kWh of power per year while avoiding the emission of about 600,000 tons of carbon dioxide into the atmosphere. Currently, the total capacity of all wind power installments in the Russian Arctic equals 210 MW. Commonly, wind power units are operated either separately or in conjunction with solar panels and diesel generators.

Solar energy plants can be installed in Eastern Yakutia. Low air temperature favors the performance of solar panels. The average annual output of solar energy in the Arctic (daytime) can reach up to 2–5 kWh and even top out at 5–6 kWh in some areas. Solar power plants operate in Republic of Sakha (Yakutia) (Batagai, Betenkes, Batamai, and other settlements) and Yamal-Nenets Autonomous District. The total solar power output equals 1.4 MW.

Hydropower plays a significant role in the energy mix in all nine territories of the Russian Arctic, especially the northwestern part of the Arctic zone. There are 17 hydropower plants in Murmansk Oblast (including the Kislogubsk tidal power plant at Ura-Guba (1.7 MW)), two in Republic of Sakha (Yakutia), and one in Arkhangelsk Oblast. The Northern power plant is being built in the Dolgaya-Vostochnaya Bay, the Kola Peninsula (12 MW, with an annual power output of 23.8 million kWh). The plant will be the first tidal power plant in Russia to reach the industrial level of energy generation. Another tidal power plant will be launched in the Mezen Bay of the White Sea (8 GW, annual power output: 38.9 billion kWh).

Arkhangelsk Oblast, Komi Republic, and other territories that have access to wood waste and industrial wood residue can produce biofuels. Conventional biofuel technologies can be improved substantially as a result of innovations. For example, innovative furnaces can supply small settlements with heat and power from wood residues with minimal heat loss, thereby producing cheap energy and waste-free wood. To date, over 50 boiler facilities in Arkhangelsk Oblast have already been converted to biofuel. Geothermal energy is also a promising sector for some Arctic territories (Butuzov, 2019). Potential reserves of heat and power in the Western Siberia artesian basin are estimated at 200 million Gcal/year.

However, renewable energy technology is still too immature to replace conventional energy sources in the foreseeable future (Aliev & Zakharcheva, 2017), especially considering the abundant deposits of oil, gas, and coal in the Russian Arctic. The total capacity of all renewable energy installations in the High North is still negligible in comparison to the extraction of hydrocarbons and the use of oil, gas, and coal in the generation of heat and electricity. The expansion of renewable energy is still limited by climate conditions and the state of modern technologies. Ensuring stable operation of a wind farm requires the use of specific materials, such as cold-resistant steel, synthetic low-temperature lubricant for bearings, and special fluids for hydraulic systems. Proper maintenance is needed, such as heating the gearbox, blades, and meteorological sensors to prevent icing. Furthermore, the solar energy output is very much dependent on seasonality. During long polar nights, solar panels are less productive. Nevertheless, reducing the carbon footprint of the energy sector by making renewable energy sources more reliable and efficient is a goal that the energy sector should aim for.

3.6.3 Small-scale power generation

The development of small-scale power-generating technologies can qualitatively transform the existing energy mix in the Russian Arctic. Such technical solutions are especially relevant for small remote settlements. Their connection to an integrated power grid is prohibitively expensive. Moreover, it requires establishing infrastructure for fuel delivery and transmission of power, which harms the environment. Thus one of the challenges to the innovation-based growth of the energy sector in the Arctic is the development of small-scale energy in the zone of decentralized power supply through more efficient use of local energy resources (Smolentsev, 2012). The small-scale energy strategy should be developed on the basis of a multifactor analysis and take into account individual climatic, economic, and social conditions in particular Arctic territories. The multifactor analysis is used to capture the territory's current and future needs in small-scale power plants and determine appropriate types of such plants. The selection of proper kinds of small-scale power plants for particular locations should be based on a comparative analysis of their competitiveness by modeling cash flows and economic efficiency parameters during the entire life cycle of the installation. The following features should be considered:

- Optimal size of territory under study
- Energy mix of territory, including future development considerations
- Energy mix of neighboring territories, including the assessment of the feasibility of import of excessive power supply from neighboring territories

- Information on existing power-generating facilities, taking into account the degree of moral and physical obsolescence and energy efficiency
- Information on energy consumers, including future consumption considerations
- Information on available energy resources (hydrocarbons, wind potential, hydropower and geothermal energy resources, solar radiation flow, etc.)

Many of small-scale power technologies have progressed to the stage at which they are being implemented in the Arctic even today. They include low-capacity nuclear power plants, small-scale power-generating installments on organic fuel (condensation, thermal, and biodiesel power plants), and renewable energy sources (wind farms, solar panels, small-scale hydropower plants, and hybrid power-generating installations) (Table 3.4).

TABLE 3.4 Small-scale power generation in the Arctic: strengths and weaknesses.

Types of small-scale generation	Strengths	Weaknesses
Low-capacity nuclear power plants	Long operation life (up to 50 years)	Extremely high construction, operation, and maintenance costs
	Autonomous operation	
	Low dependence on fuel deliveries and other supplies	Long payback period
	Neutral environmental effect	
Small-scale power-generating installments on organic fuel	Low start-up costs	Dependence on fuel supply
	Accessible technology, no need for high-tech equipment	High operation costs due to cost of fuel delivered from other territories or transported over long distances
		Adverse environmental impact
Other renewable sources of power	Relatively low start-up costs	High operation and maintenance costs
	Low or neutral environmental effect	Overexposure to weather conditions and climate
	No dependence on fuel deliveries	Unstable generation

Authors' development based on Smolentsev, D. (2012). Energy development in the Arctic: challenges Challenges and opportunities of low-power plants. Arctic: Ecology and Economy, *3(7), 22–29.*

Low-capacity nuclear power plants are particularly relevant for the Arctic, owing to their reliability, independence from fuel supply, and neutral environmental impact. Existing innovative solutions in the sphere of small-scale nuclear power are characterized by a high degree of adaptability of technical, economic, and operational features. These include location alternatives (underground, inland, floating) and a range of capacity characteristics (from low-capacity installations to large complexes). Major competitive advantages of low-capacity nuclear power plants include flexibility (the module layout allows for creating structures of any capacity), long-term operation without refueling, mobility, and relatively low cost of generated power. An example is the world's first Arctic floating nuclear power plant Akademik Lomonosov, the flagship in a series of mobile transportable small-scale power units (Khvostova, 2018). It is located in Pevek, Chukotka Autonomous District. The plant ensures a reliable year-round heat and power supply to remote territories in Chukotka and the Far East. Being connected to the isolated Chaun-Bilibino power grid, it replaces the retired capacities of the Bilibino nuclear power station and the Chaun cogeneration heat and power plant. The floating nuclear plant also supplies energy to mining companies in western Chukotka in the Chaun-Bilibino energy hub—a large ore, metal, and gold cluster in the Baim ore zone. Designed for an operational life of 40 years with an extension of 10 years, the Akademik Lomonosov facility is expected to generate 50 Gcal/h of heat energy, which is an equivalent of the power supply for more than 100,000 households per year (NS Energy, 2021). Thus along with other types of small-scale power generation, low-capacity nuclear power plants can contribute to solving the energy supply problem across the Russian Arctic and establishing a stable system of reliable, independent, and low-carbon regional and local power supply.

3.6.4 Niches for international collaboration

Among the priorities for the innovative development of the energy sector, Russia has declared the need to expand international collaboration in the development and implementation of innovative technologies through the following measures (Government of the Russian Federation, 2020b):

- Enhancing participation in international efforts to ensure sustainable global energy development, including the United Nations Sustainable Development Goals (United Nations, 2021c) (notably, Goal 7, "Ensure Access to Affordable, Reliable, Sustainable and Modern Energy for All" (United Nations, 2021a))
- Establishing common energy markets within the Eurasian Economic Union (oil and petroleum products, natural gas, and electricity) on the principles of fair competition; free movement of goods, services,

technologies, and investments in the energy sector; and coordination of national policies in the sphere of mineral resources management
- Strengthening cooperation with the Organization of the Petroleum Exporting Countries, the Gas Exporting Countries Forum, other international multilateral organizations in the energy sector, and individual energy-exporting countries
- Participation in international negotiations on energy issues, including the climate-related agenda and the pursuance of the United Nations Sustainable Development Goal 13, "Take Urgent Action to Combat Climate Change and Its Impacts" (United Nations, 2021b);
- Strengthening the legal framework for energy cooperation to balance the interests of exporters and importers of energy resources, as well as of transit countries
- Building and promotion of a positive image of the Russian energy sector

However, with the introduction of international sanctions against Russia, cooperation between individual companies has changed radically. Most U.S. and European companies have withdrawn from joint projects with their Russian counterparts in the Arctic or have cut off equipment and technologies supplies to Russia. This situation forces Russia to replace the resulting technological gaps with domestic equipment (import substitution policy) or to switch to alternative suppliers (e.g., China) (Gao & Erokhin, 2020a). Thus new niches in the market are opening up to be filled by developers and producers of various types of energy equipment, materials, and technological solutions from non-Arctic states. The most demanded technologies and equipment can be divided into three groups according to the dominant segments of the Arctic energy sector: oil and gas, coal, and electricity (Table 3.5).

Many scholars (Avramenko & Baiguskarova, 2018; Dun & Lukin, 2018; Huang & Korolev, 2015; Wang, 2020) advocate collaboration in the energy sector with emerging and technologically advanced non-Arctic countries, such as China, Japan, and South Korea. China and Japan already participate in joint energy projects with Russia (Yamal LNG and Arctic LNG), while South Korea builds LNG tankers for Novatek. Although many of the developments that are available in the Asian market are not adapted to the specific conditions of the Arctic, this gap can be relatively quickly eliminated by testing various technological solutions and improving technical characteristics to the desired level. In particular, China is a world leader in developing renewable energy (particularly solar energy and wind farms). Moreover, China is interested in creating additional energy generation for domestic needs, making Russia's Far East and the Arctic particularly attractive to Chinese investors. China's Polar Silk Road initiative envisages the development

TABLE 3.5 Technologies and equipment Russia demands in the energy sector.

Sectors	Technologies and equipment
Oil and gas	Rock properties analysis, including boundary effects and phase transformations
	Offshore and onshore seismic surveys, including automatic processing and interpretation of seismic data in stream mode and wireless data acquisition systems
	Geoinformation systems
	Analysis of rock and reservoir fluid, including studying wells with a sound depth of several meters and telemetry for data transmission from downhole equipment
	Reservoir stimulation, including hydraulic fracturing
	Smart field equipment and technologies, including pumps and flowmeters for multiphase flow
	Automated control and monitoring of technological processes and equipment
	Development of hard-to-recover reserves, including directional drilling
	In situ conversion, including kerogen conversion techniques and technologies
	Equipment and technologies for natural gas liquefaction
	Processing hydrocarbons, including additives to oil and fuels and catalysts for oil refining
	Design and construction of large-capacity modules for fuel and energy facilities
	Development of offshore fields, including subsea production units
	Ice-class drilling complexes
Coal	Real-time monitoring of parameters and operation modes of equipment
	Monitoring and control of parameters of rock massifs
	Coal chemistry for the production of higher added value goods, including liquid fuels
	Prevention of spontaneous combustion of coal in rock massifs and coal storages
	Stripping equipment and coal mining and quarry transport equipment (hydraulic excavators, tunneling combines, high-performance miner-bolters, control systems for mechanized hydraulic supports for underground coal mining, automated mining complexes and units)
	Coal-cleaning equipment and technologies
	Automated transport vehicles, including hydraulic transportation of coal
	Shearers for high coal deposits
	Explosion-proof electric engines
	Dust suppression and dust-explosion safety systems integrated into tunneling and cleaning equipment

(Continued)

TABLE 3.5 (Continued)

Sectors	Technologies and equipment
Electricity and renewable energy	Supercritical parameters of steam with improved technical, economic, and environmental features
	Higher performance of turbines through changes in their parameters and the use of new working fluids, including carbon dioxide
	Electrical energy accumulation systems, including batteries and fuel cells
	Automated control and monitoring of technological processes and equipment, intelligent power networks, and digital information transmission devices
	Power gas turbines (65 MW and higher)
	Gas-insulated electrical equipment
	High-capacity wind turbines
	High-voltage and generator circuit breakers
	Electrical equipment with semiconductor elements (static reactive power compensators, inverters, converters, rectifiers)
	High-performance photovoltaic modules
	Digital twins, including means of conducting comprehensive digital tests of equipment and technologies and confirming reliability parameters
	Autonomous power generators on gaseous and hydrogen fuel and other autonomous sources of energy supply designed for continuous power generation

Authors' development based on Government of the Russian Federation. (2020b). Decree #1523-r from June 9, 2020. Energy strategy of the Russian Federation for the period until 2035. *<https://minenergo.gov.ru/node/1026>; Accessed 08.12.21.*

of transport and economic corridor running through the Arctic Ocean to Europe and North America (Erokhin & Gao, 2021; Erokhin, Gao, & Zhang, 2021). To make this vision come true, Russia's collaboration with foreign partners in the Arctic should advance from spotty resource projects (Yamal LNG, Arctic 2 LNG, etc.) to comprehensive cooperation in the spheres of spatial development, investment, trade, and innovations (Gao & Erokhin, 2021). Therefore collaboration with China and other countries in diversification of the energy mix could contribute to the sustainable development of circumpolar territories.

3.7 Conclusion

Achieving the UN Sustainable Development Goals in the field of energy and climate in the Arctic is an extraordinarily complex and multifaceted task. It requires a radical reduction in emissions and an increase in energy efficiency across all sectors of the Arctic economy (the energy sector itself, industrial production, transport, shipping, construction, agriculture, forestry, and municipal services). In parallel, it involves a radical change in the approach to the development and use of energy resources (environmental management, energy conservation, transition to renewable and green energy). Russia enjoys enormous potential in Arctic energy. Until recently, the energy sector's growth in the Russian High North, Siberia, and the Far East has been spurred by conventional energy reserves, such as coal, oil, and natural gas. The study shows that the development of energy-related innovative technologies has opened up opportunities for other types of energy generation, including LNG, small-scale nuclear energy, hydropower, and renewable wind, solar, and biofuel energy. However, no one country can meet a sustainable development challenge alone. Effective implementation of modern innovative solutions requires comprehensive cooperation among the world's leading developers, producers, and suppliers of technologies and equipment in the Arctic energy sector.

Acknowledgments

The study is supported by the National Key R&D Program of China (project no. 2020YFE0202600) and the Grant of Central Universities of the Ministry of Education of the People's Republic of China.

References

Aliev, R., & Avramenko, A. (2017). *Environmental problems of global fuel and energy sector*. Moscow: Moscow MGIMO University.

Aliev, R., & Zakharcheva, K. (2017). Changes in the ratio of generating capacities in different countries: From fossil fuels to alternative energy. *Bulletin of the Moscow Region State University*, 2, 8–17.

Andrei, J. V., Avram, S., Băncescu, I., Gâf Deac, I. I., & Gheorghe, C. (2021). Evolution of Romania's economic structure and environment degradation – An assessment through LMDI decomposition approach. *Risk Management and Healthcare Policy*, *14*, 3505–3521.

Armeanu, D. S., Joldes, C. C., Gherghina, S. C., & Andrei, J. V. (2021). Understanding the multidimensional linkages among renewable energy, pollution, economic growth and urbanization in contemporary economies: Quantitative assessments across different income countries' groups. *Renewable and Sustainable Energy Reviews*, *142*, 110818.

Assembly of Deputies of Arkhangelsk Oblast. (2019). Decree #57-5-OZ from February 18, 2019. *On the approval of the strategy of social and economic development of Arkhangelsk Oblast till 2035*. <https://docs.cntd.ru/document/462642024>; Accessed 07.12.21.

Assembly of Deputies of Nenets Autonomous District. (2019). Decree #256-sd from November 7, 2019. *On the approval of the strategy of social and economic development of nenets autonomous district till 2030*. <https://docs.cntd.ru/document/561620008>; Accessed 07.12.21.

Avramenko, A., & Baiguskarova, A. (2018). Collaboration between China and Russia in the sphere of use of renewable sources of energy. *Eurasian Union of Scientists*, *49*(4), 4–6.

Balakireva, A., Dubinin, A., Kormishin, A., Yaresko, A., Zakharova, A., Shiptenko, A., & Timofeeva, I. (2018). *Arctic 18–24-35: Youth outlook*. Arkhangelsk: Center for Strategic Development.

Bansal, R. C., Zobaa, A. F., & Saket, R. K. (2005). Some issues related to power generation using wind energy conversion systems: An overview. *International Journal of Emerging Electric Power Systems, 3*(2), 1070.

Berdin, V., Kokorin, A., Yulkin, G., & Yulkin, M. (2017). *Renewable energy sources in remote settlements in the Russian Arctic*. Moscow: World Wildlife Fund.

Brekhuntsov, A., Petrov, Y., & Prykova, O. (2020). Ecological aspects of the development of the natural resource potential of the Russian. *Arctic: Ecology and Economy, 3*, 34–47.

Butuzov, V. (2019). Russian geothermal energy: Analysis of hundred development of scientific and engineering concepts. *Journal of Environmental Earth and Energy Study, 3*, 4–21.

Cai, Q., Wang, J., Beletsky, D., Overland, J., Ikeda, M., & Wan, L. (2021). Accelerated decline of summer Arctic sea ice during 1850–2017 and the amplified Arctic warming during the recent decades. *Environmental Research Letters, 16*(3), 034015.

Collins, B., & Kumral, M. (2020). Environmental sustainability, decision-making, and management for mineral development in the Canadian Arctic. *International Journal of Sustainable Development and World Ecology, 27*(4), 297–309.

Danubianu, M., & Teodorescu, C. (2021). Energy, climate changes, covid-19: Romanian answers. *Present Environment and Sustainable Development, 15*(1), 59–67.

Dincer, I. (2000). Renewable energy and sustainable development: A crucial review. *Renewable and Sustainable Energy Reviews, 4*(2), 157–175.

Dun, T., & Lukin, A. (2018). The prospects for Russia-China collaboration in the Arctic and the Russian far east. *Ojkumena. Regional Researches, 4*, 158–167.

Ellabban, O., Abu-Rub, H., & Blaabjerg, F. (2014). Renewable energy resources: Current status, future prospects and their enabling technology. *Renewable and Sustainable Energy Reviews, 39*, 748–764.

Erokhin, V., & Gao, T. (2021). Better connectivity for greater Eurasia: China-Russia cooperation on the development of the economic corridor in the high north. In A. Aseeva, & J. Gorski (Eds.), *The law and policy of new Eurasian regionalization: Economic integration, trade, and investment in the post-soviet and greater Eurasian space* (pp. 32–47). Leiden: Brill Nijhoff.

Erokhin, V., Gao, T., & Zhang, X. (2021). Arctic blue economic corridor: China's role in the development of a new connectivity paradigm in the north. In J. Barnes, H. Exner-Pirot, L. Heininen, & P. W. Lackenbauer (Eds.), *China's Arctic engagement: Following the polar silk road to Greenland and Russia* (pp. 63–90). Peterborough: North American and Arctic Defence and Security Network.

Gao, T., & Erokhin, V. (2020a). China-russia collaboration in Arctic shipping and maritime engineering. *The Polar Journal, 10*(2), 353–374.

Gao, T., & Erokhin, V. (2020b). Widening the scope of responses to environmental concerns in the high north: Arctic countries' policies and the role of China. In L. Heininen, H. Exner-Pirot, & J. Barnes (Eds.), *Climate change and the Arctic: Global origins, regional responsibilities? Arctic yearbook 2020* (pp. 357–397). Akureyri: Arctic Portal.

Gao, T., & Erokhin, V. (2021). How Russia's new vision of territorial development in the arctic can boost china-russia economic collaboration. In L. Heininen, H. Exner-Pirot, & J. Barnes (Eds.), *Climate change and the Arctic: Global origins, regional responsibilities? Arctic yearbook 2021* (pp. 249–278). Akureyri: Arctic Portal.

Gao, T., Bobylev, N., Gadal, S., Lagutina, M., Sergunin, A., & Erokhin, V. (2021). Planning for sustainability: An emerging blue economy in Russia's coastal Arctic? *Sustainability, 13*(9), 4957.

Gernaat, D. E. H. J., de Boer, H. S., Daioglou, V., Yalew, S., Muller, C., & van Vuuren, D. (2021). Climate change impacts on renewable energy supply. *Nature Climate Change, 11*, 119–125.

Gerstlberger, W., Knudsen, M. P., Dachs, B., & Schroter, M. (2016). Closing the energy-efficiency technology gap in european firms? Innovation and adoption of energy efficiency technologies. *Journal of Engineering and Technology Management, 40*, 87–100.

Government of Chukotka Autonomous District. (2014). Decree #290-rp from July 16, 2014. *Strategy of social and economic development of Chukotka Autonomous District till 2030*. <https://docs.cntd.ru/document/446123709>; Accessed 07.12.21

Government of Krasnoyarsk Krai. (2020). Decree #122-r from February 26, 2020. *On the approval of the strategy of social and economic development of Northern and Arctic territories and support of indigenous people in Krasnoyarsk Krai till 2030*. <https://docs.cntd.ru/document/570728130>; Accessed 07.12.21.

Government of Murmansk Oblast. (2013). Decree #768-PP/20 from December 25, 2013. *Strategy of social and economic development of Murmansk Oblast till 2020 and for the period until 2025*. <https://minec.gov-murman.ru/activities/strat_plan/sub02/>; Accessed 07.12.21.

Government of the Komi Republic. (2019). Decree #185 from April 11, 2019. *On the strategy of social and economic development of the Komi Republic till 2035*.<https://docs.cntd.ru/document/553237768>; Accessed 07.12.21.

Government of the Republic of Karelia. (2018). Decree #899r-P from December 29, 2018. *On the approval of the strategy of social and economic development of the republic of Karelia till 2030*. <https://docs.cntd.ru/document/465420565>; Accessed 07.12.21.

Government of the Russian Federation. (2009). Decree #1-p from January 8, 2009, (updated on June 1, 2021). *On the approval of the main directions of the state policy in the field of improving the energy efficiency of the energy sector based on the use of renewable energy for the period until 2035*. <http://government.ru/docs/all/66930/>; Accessed 07.12.21.

Government of the Russian Federation. (2014). Decree #321 from April 15, 2014. *On the approval of the state program of the Russian Federation on energy efficiency and development of the energy sector*. <http://government.ru/docs/all/91334/>; Accessed 07.12.21.

Government of the Russian Federation. (2020a). Decree #973-r from April 10, 2020. *On the approval of the individual program of social and economic development of the Republic of Karelia for 2020–2024*. <https://docs.cntd.ru/document/564652911?marker = 65A0IQ>; Accessed 07.12.21.

Government of the Russian Federation. (2020b). Decree #1523-r from June 9, 2020, "Energy Strategy of the Russian Federation for the Period until 2035." <https://minenergo.gov.ru/node/1026>; Accessed 08.12.21.

Government of the Russian Federation. (2021). Decree #484 from March 30, 2021. *On the approval of the state program of social and economic development of the Arctic zone of the Russian Federation*. <https://docs.cntd.ru/document/603154509?marker = 6540IN>; Accessed 07.12.21.

Governor of Murmansk Oblast. (2014). Decree #126-PG from August 27, 2014. *On approval of the regional program of social and economic development "Murmansk Oblast – Strategic Center of the Arctic Zone of the Russian Federation"*. <https://docs.cntd.ru/document/412386353>; Accessed 07.12.21.

Governor of Yamal-Nenets Autonomous District. (2018). *Draft strategy of social and economic development of Yamal-Nenets Autonomous District till 2030*. <https://www.yanao.ru/documents/all/57256/>; Accessed 07.12.21.

Head of the Republic of Sakha (Yakutia). (2020). Decree #1377 from August 14, 2020. *On the Strategy of social and economic development of the Arctic zone of the Republic of Sakha (Yakutia) till 2035*. <https://www.sakha.gov.ru/news/front/view/id/3204989>; Accessed 07.12.21.

Huang, J., & Korolev, A. (2015). *International cooperation in the development of Russia's far east and Siberia*. London: Palgrave Macmillan.

Huntington, H., Boyle, M., Flowers, G., Weatherly, J., Hamilton, L., Hinzman, L., & Overpeck, J. (2007). The influence of human activity in the Arctic on climate and climate impacts. *Climatic Change, 82*(1–2), 77–92.

Irandoust, M. (2019). On the causality between energy efficiency and technological innovations: Limitations and implications. *International Journal of Green Energy, 16*(15), 1665–1675.

Jessoe, K., & Moore, F. (2021). The energy costs of climate change. *Nature, 598*(7880), 262–263.

Khambalkar, V. P., Katkhede, S. S., Dahatonde, S., Korpe, N. D., & Nage, S. M. (2010). Renewable energy: An assessment of public awareness. *International Journal of Ambient Energy, 31*(3), 133–142.

Khvostova, M. (2018). Environmental problems of operation of floating nuclear power plant in the Arctic region. *Russian Arctic, 1*, 12–29.

Kiushkina, V., Samsonov, R., Ishmuratova, M., Duskabilova, Z., Rodichkin, I., Sung, J., & Chuvychkina, I. (2020). *Arctic strategies: Energy, security, ecology, and climate*. Moscow: Moscow School of Management Skolkovo.

Klimenko, V., Klimenko, A., Tereshin, A., & Mikushina, O. (2021). Energy and natural climate factors: If there is a chance to stop global warming? *Energy Policy, 158*(4), 12–29.

Kokorin, A. (2017). Climate change as basis of the paris agreement. *Energy Policy, 3*, 34–37.

Kung, C., & McCarl, B. (2018). Sustainable energy development under climate change. *Sustainability, 10*(9), 3269.

Legislative Assembly of Yamal-Nenets Autonomous District. (2011). Decree #839 from December 14, 2011. *On the strategy of social and economic development of Yamal-Nenets Autonomous District till 2020*. <https://docs.cntd.ru/document/424064239>; Accessed 07.12.21.

Lundy, M. (2019). Renewable energy goals in the face of climate change. *Berkeley Scientific Journal, 23*(2), 18–21.

Luoto, T., Ojala, A., & Zajaczkowski, M. (2019). Proxy-based 300-year high Arctic climate warming record from Svalbard. *Polar Record, 55*(3), 132–141.

Maiorov, M. (2020). *Alternative energy in the Arctic*. <https://goarctic.ru/work/alternativnaya-energetika-v-arktike/>; Accessed 11.12.21.

Makarov, I., & Stepanov, I. (2015). Environmental factor of economic development of Russian Arctic. *EKO, 11*, 120–138.

McCrystall, M., Stroeve, J., Serreze, M., Forbes, B., & Screen, J. (2021). New climate models reveal faster and larger increases in arctic precipitation than previously projected. *Nature Communications, 12*(1), 6765.

Medvedeva, O. (2015). The tasks of the assessment of environmental damage in the Arctic. *Arctic and North, 18*, 131–147.

Melnik, A., Naoumova, I., Ermolaev, K., & Katrichis, J. (2021). Driving innovation through energy efficiency: A Russian regional analysis. *Sustainability, 13*(9), 4810.

Miao, C., Fang, D., Sun, L., Luo, Q., & Yu, Q. (2018). Driving effect of technology innovation on energy utilization efficiency in strategic emerging industries. *Journal of Cleaner Production, 170*, 1177–1184.

Morgunova, M., & Kovalenko, A. (2021). Energy innovation in the Arctic. *Energy Policy, 158*(4), 30–43.

Muhammed, G., & Tekbiyik-Ersoy, N. (2020). Development of renewable energy in China, USA, and Brazil: A comparative study on renewable energy policies. *Sustainability, 12*, 9136.

Neacsa, A., Panait, M., Muresan, J. D., & Voica, M. C. (2020). Energy poverty in european union: Assessment difficulties, effects on the quality of life, mitigation measures. Some evidences from Romania. *Sustainability, 12*(10), 4036.

Nosko, P. (2017). The paris climate agreement as a determinant of the future global economy and its implications for russia. *Bulletin of the Financial University, 1*, 145–150.

Novatek. (2021a). *Arctic LNG 2*. <http://www.novatek.ru/en/business/arctic-lng/>; Accessed 11.12.21.

Novatek. (2021b). *Yamal LNG Infrastructure.* <http://www.novatek.ru/en/business/yamal-lng/yamal_infrastructure/>; Accessed 11.12.21.

NS Energy. (2021). *Akademik Lomonosov Floating Nuclear Power Plant.* <https://www.nsenergybusiness.com/projects/akademik-lomonosov-floating-npp/>; Accessed 14.12.21.

Panait, M., Voica, M. C., & Radulescu, I. G. (2019). Approaches regarding environmental kuznets curve in the european union from the perspective of sustainable development. *Applied Ecology and Environmental Research, 17*(3), 6801–6820.

Polvani, L., Previdi, M., England, M., Chiodo, G., & Smith, K. (2020). Substantial twentieth-century Arctic warming caused by ozone-depleting substances. *Nature Climate Change, 10*(2), 130–133.

Popescu, G. H., Andrei, J. V., Nica, E., Mieilă, M., & Panait, M. (2019). Analysis on the impact of investments, energy use and domestic material consumption in changing the Romanian economic paradigm. *Technological and Economic Development of Economy, 25* (1), 59–81.

President of the Russian Federation. (2020a). Decree #164 from March 5, 2020. *On the foundations of the state policy of the Russian Federation in the Arctic till 2035.* <http://www.kremlin.ru/acts/news/62947>; Accessed 07.12.21.

President of the Russian Federation. (2020b). Decree #645 from October 26, 2020. *On the strategy for the development of the Arctic zone of the Russian Federation and ensurance of national security till 2035.* <https://docs.cntd.ru/document/566091182>; Accessed 07.12.21.

Sayigh, A. (2021). Up-date: Renewable energy and climate change. *Renewable Energy and Environmental Sustainability, 6*, 13.

Smolentsev, D. (2012). Energy development in the Arctic: Challenges and opportunities of low-power plants. *Arctic: Ecology and Economy, 3*(7), 22–29.

Song, Z., & Zhang, N. (2020). Research on green innovation efficiency of listed Chinese energy companies based on triple bottom line. *Complexity, 2020*, 3450471.

Streimikiene, D., Balezentis, T., & Krisciukaitiene, I. (2012). Promoting Interactions between Local Climate Change Mitigation, Sustainable Energy Development, and Rural Development Policies in Lithuania. *Energy Policy, 50*, 699–710.

Taseska, V., Markovska, N., & Callaway, J. (2012). Evaluation of climate change impacts on energy demand. *Energy, 48*(1), 88–95.

Toke, D., & Vezirgiannidou, S. (2013). The relationship between climate change and energy security: Key issues and conclusions. *Environmental Politics, 22*(4), 537–552.

Trump, B., Kadenic, M., & Linkov, I. (2018). A sustainable Arctic: Making hard decisions. *Arctic, Antarctic, and Alpine Research, 50*(1), e1438345.

United Nations Development Programme. (2021). *The SDGs in action.* <https://www.undp.org/sustainable-development-goals>; Accessed 14.12.21.

United Nations Framework Convention on Climate Change. (2018). *UN climate change annual report 2018*. Bonn: United Nations Framework Convention on Climate Change.

United Nations. (2001). *Stockholm Convention on Persistent Organic Pollutants (POPs).* <https://www.un.org/ldcportal/stockholm-convention-on-persistent-organic-pollutants-pops/>; Accessed 13.12.21.

United Nations. (2015). *The Paris Agreement.* <https://www.un.org/en/climatechange/paris-agreement>; Accessed 15.12.21.

United Nations. (2021a). *Goal 7: Ensure access to affordable, reliable, sustainable and modern energy for all.* <https://sdgs.un.org/goals/goal7>; Accessed 09.12.21.

United Nations. (2021b). *Goal 13: Take urgent action to combat climate change and its impacts.* <https://sdgs.un.org/goals/goal13>; Accessed 09.12.21.

United Nations. (2021c). *The 17 Goals.* <https://sdgs.un.org/goals>; Accessed 09.12.21.

Urpelainen, J. (2021). Export orientation and domestic electricity generation: Effects on energy efficiency innovation in select sectors. *Energy Policy, 39*(9), 5638–5646.

Van, L. (2014). Green growth: The sustainable economic growth model. *Science and Technology Development Journal*, *17*(3), 6–13.

Wang, Y. (2020). The status and challenges of sino-Russian far east cooperation. *American Scientific Journal*, *43*, 12–15.

Xu, Z., Assenova, A., & Erokhin, V. (2018). Renewable energy and sustainable development in a resource-abundant country: Challenges of wind power generation in Kazakhstan. *Sustainability*, *10*(9), 3315.

Zhilina, I. (2021). Warming in the Arctic: Opportunities and risks. *Economic and Social Problems of Russia*, *1*, 66–87.

CHAPTER

4

Analysis of factors influencing smart and sustainable manufacturing systems using a multi-criteria decision making tool

Anilkumar Malaga and S. Vinodh

Department of Production Engineering, National Institute of Technology, Tiruchirappalli, Tamil Nadu, India

4.1 Introduction

The manufacturing industries face issues resulting from the increase in competition due to the enhancements and restraints such as quality and cost, especially from developing countries (Barrientos & Gereffi, 2011). Simultaneously, there has been an acceleration in data availability, which is connected to the latest technologies and provides a foundation for smart and sustainable enterprise (Edwards, 2021) and the possibility for the desired optimization of productive systems. Process, equipment, and quality (Singh, Khamba, & Singh, 2021) are essential areas in which to optimize overall equipment effectiveness (OEE). OEE involves efficiency, availability, and quality are sometimes used to connect those (Malaga & Vinodh, 2021a). Technologies varying from monitoring and regulation to analytics and decision support are being developed to optimize the information and expert knowledge (Wang, Hu, Sun, & Freiheit, 2020). In the present scenario there is an enhancement in interest in analytics and

Innovation and Sustainable Manufacturing
DOI: https://doi.org/10.1016/B978-0-12-819513-0.00002-0

decision support technologies of Industry 4.0 (Guo, Wang, Xu, & Wu, 2020), such as machine learning and artificial intelligence (AI), which are appropriate for interpreting detailed data, knowledge, and adaptability (Susana Ferreiro et al., 2021).

Industry 4.0 technologies are frequently used to boost competitiveness by enhancing the robustness and reliability of production in a smart manufacturing (SM) context. The automobile component manufacturing industry is one of the most competitive (Dib, Ribeiro, & Prates, 2021), with innovations and reduction in costs being crucial to attaining their objectives. Initiatives are being pushed toward the notion of SM (Malaga & Vinodh, 2021b) to successfully implement in manufacturing industries and generate a mark on the global index (Suvarna et al., 2020). On the other side, industrialization and fast development in manufacturing processes in industries have resulted in severe levels of environmental damage (Yang, Usman, & Jahanger, 2021). The causes might range from poor industrial and legislative strategies to overcrowding and uncontrolled urbanization (Isharyani et al., 2021). As a result, industries need to actively implement sustainable manufacturing and SM techniques (Lenz, MacDonald, Harik, & Wuest, 2020). However, this sustainable smart (SSM) concept is essential to increasing current production capacity to satisfy rising demand. It is unclear whether the two concepts of SM and sustainability will interact positively or negatively (Abubakr et al., 2020). To adopt SSM in the industry, there is a need to analyze the factors that will influence smart and sustainable manufacturing systems. To fill the research gap, the authors identified the factors influencing the adoption of SSM through a literature analysis that was validated by the expert panel. Then, identified factors were analyzed by using a multicriteria decision-making tool known as fuzzy Technique for Order of Preference by Similarity to Ideal Solution (TOPSIS).

This work is organized as follows: Section 4.2 details the literature review, a case study is presented in Section 4.3, and Section 4.4 presents the conclusions of the study.

4.2 Literature review

Shukla, Jangid, Siddh, Kumar, and Soni (2017) evaluated the important variables of sustainable production in Indian car manufacturing through empirical research that uses an analytic hierarchy process. The authors utilized a survey approach with both academic and industry involvement. The respective relevance of the elements for sustainability was derived through brainstorming process and a survey that identified the evaluation of the factors in three particular firms. The authors concluded with the development of a rank analysis of the firms. The first car manufacturer received the highest final grade, indicating that it effectively adheres to sustainable manufacturing principles.

Chavarría-Barrientos, Batres, Wright, and Molina (2018) proposed a technique for designing a sensing, smart, and sustainable manufacturing industry based on enterprise architecture concepts. The method strives to implement best practices in enterprise engineering while filling any gaps that may exist. The proposed technique covers critical issues to consider in incorporating S3 traits into enterprise architecture. The authors showed that the proposed technique incorporates management-level perspectives, which is especially crucial in the early stages of design. The enterprise viewpoint specifies the business perspective through the establishment of strategies and business models.

Abubakr et al. (2020) proposed an integrated approach to incorporating sustainable SM performance by embracing sustainable manufacturing measures and evaluating existing and future manufacturing sector issues. The authors determined that the SSM model should be built on particular policies to improve competence to fulfill immediate infrastructure requirements. This method will enhance environmental quality for subsequent generations and future business projections for production industries. The authors also facilitated future research prospects involving sustainable SM.

Raman et al. (2020) presented a review of existing literature that indicates research possibilities in various modern manufacturing sectors. Possible directions for future investigation outline the modern manufacturing community's short-, mid-, and long-term demands for achieving smart and sustainable production. The authors found that AI and other evolving technologies can significantly affect the process- and system-level developments among the production sectors, and enhanced production education may motivate future generations to work in production engineering and research.

Susana Ferreiro et al. (2021) proposed a technique in the additive manufacturing process and its relevance in a laser-cladding situation. The authors explored the present state of analytics approaches that had not been changed in years. A correlation study concentrates on data analysis to derive interrelations between process data and process parameters corresponding to the quality of the test output. Consequently, the authors showed that the porosity was the most difficult to create and model for the two quality indicators that were tested. The model yielded modest findings, although the outcomes for dimensionality were better, and many factors demonstrated relevance for correlations.

Aggarwal et al. (2021) discovered a relationship between sustainable and SM methods in developing economies. The authors employed hypothesis modeling to relate production competitiveness and top leadership emphasis on sustainable and SM activities. The authors gathered the information through a questionnaire survey that was administered in the Indian industrial sectors. The authors focused on how manufacturing industries in emerging economies such as India may implement SSM techniques and validated the relations using the structural equation modeling technique.

4.3 Case study

The objective of this study is to investigate the factors influencing SSM adoption using an appropriate methodology. Various factors have been identified that influence the adoption of SSM systems in the case industry. Suitable criteria were defined, and significant factors can be investigated by using fuzzy TOPSIS. The inputs for the investigation were obtained through experts, and further computations were carried out by using fuzzy TOPSIS methodology to prioritize the influencing factors.

The step-by-step investigation process using fuzzy TOPSIS for this study is as follows (Vinodh, Thiagarajan, & Mulanjur, 2014):

Step 1: Identify the suitable criteria and appropriate factors influencing the adoption of SSM systems.

Appropriate factors influencing the adoption of SSM systems were identified from the literature.

Table 4.1 presents the list of factors influencing the adoption of SSM systems. In total, 14 factors were identified as influencing factors for the adoption of SSM systems. The identified 14 factors were investigated under five criteria: innovation (I), technology readiness (TR), flexibility (F), customization (C), and return on investment (RoI).

Step 2: Identify the suitable members for the expert panel inclusive of academicians and industry practitioners.

An expert panel was established with experts from academia or research and development and industry to facilitate inputs for the investigation purpose. The identified factors and criteria were validated by the expert panel. To facilitate the inputs for the investigation, the experts used a suitable scale.

Step 3: Identify the appropriate linguistic terms on a suitable scale in order to gather inputs for the criteria weights and factor ratings from the expert panel.

The linguistic terms of factor ratings can be transformed into fuzzy numbers using a seven-point scale (Vinodh et al., 2014), consisting of Very Poor (VP)—(0, 0, 1, 2); Poor (P)—(1, 2, 2, 3); Medium Poor (MP)—(2, 3, 4, 5); Fair (F)—(4, 5, 5, 6); Medium Good (MG)—(5, 6, 7, 8); Good (G)—(7, 8, 8, 9); and Very Good (VG)—(8, 9, 10, 10). The scale that was used for criteria weights consisted of Very Low (VL)—(0, 0, 0.1, 0.2);

TABLE 4.1 Factors influencing the adoption of smart and sustainable manufacturing systems.

S. no.	Factors	Authors
1	Top management commitment	Aggarwal et al. (2021), Agrawal, Wankhede, Kumar, Upadhyay, and Garza-Reyes (2021)
2	Security	Meng, Yang, Chung, Lee, and Shao (2018)
3	Economy	Chavarría-Barrientos et al. (2018), Abubakr et al. (2020), Aggarwal et al. (2021)
4	Integration	Chavarría-Barrientos et al. (2018), Raman et al. (2020), Aggarwal et al. (2021)
5	Reliability	Raman et al. (2020), Aggarwal et al. (2021), Dib et al. (2021)
6	Automation	Raman et al. (2020), Molina, Ponce, Miranda, and Cortés (2021)
7	Increased productivity	Raman et al. (2020), Molina et al. (2021)
8	3R and 6R adoption	Agrawal et al. (2021), Molina et al. (2021)
9	Energy consumption	Raman et al. (2020), Molina et al. (2021)
10	Workforce training	Dib et al. (2021), Raman et al. (2020), Susana Ferreiro et al. (2021)
11	Compliance to regulations	Chavarría-Barrientos et al. (2018), Abubakr et al. (2020)
12	Resource utilization	Abubakr et al. (2020), Suvarna et al. (2020), Wu et al. (2022)
13	Social responsibility	Chavarría-Barrientos et al. (2018), Raman et al. (2020), Agrawal et al. (2021)
14	Technology adoption	Sunmola and Apeji (2020), Lepore, Micozzi, and Spigarelli (2021)

Low (L)—(0.1, 0.2, 0.2, 0.3); Medium Low (ML)—(0.2, 0.3, 0.4, 0.5); Medium (M)—(0.4, 0.5, 0.5, 0.6); Medium High (MH)—(0.5, 0.6, 0.7, 0.8); High (H)—(0.7, 0.8, 0.8, 0.9); and Very High (VH)—(0.8, 0.9, 1, 1).

Step 4: The expert panel was requested to provide factor ratings and criteria weights as per the scale in linguistic terms.

The collected factor ratings and criteria weights from the expert panel are listed in Table 4.2.

Table 4.2 represents inputs gathered from the expert panel individually for every factor, and the expert panel provided a consensus opinion for the criteria weights.

TABLE 4.2 Collected Inputs from expert panel for factor ratings and criteria weights.

S. no	Factors	Criteria									
		Innovation (I)		Technology Readiness (TR)		Flexibility (F)		Customization (C)		Return on Investment (RoI)	
		DM1	DM2	DM1	DM2	DM1	DM2	DM1	DM2	DM1	DM2
1	Top management commitment ($\mathscr{F}1$)	G	MG	G	MG	G	VG	G	MG	G	MG
2	Security ($\mathscr{F}2$)	MG	F	G	MG	MG	F	F	F	G	F
3	Economy ($\mathscr{F}3$)	G	VG	G	MG	G	F	F	MG	G	MG
4	Integration ($\mathscr{F}4$)	G	MG	G	VG	VG	G	G	MG	G	MG
5	Reliability ($\mathscr{F}5$)	MG	G	G	MG	G	G	MG	G	G	G
6	Automation ($\mathscr{F}6$)	MG	MG	G	G	G	MG	MG	G	G	G
7	Increased productivity ($\mathscr{F}7$)	G	MG	G	MG	G	G	G	MG	MG	MG
8	3R and 6R adoption ($\mathscr{F}8$)	G	MG	MG	G	MG	G	MG	F	G	MG
9	Energy consumption ($\mathscr{F}9$)	MG	G	G	MG	MG	F	F	MG	G	MG
10	Workforce training ($\mathscr{F}10$)	G	MG	G	MG	F	MG	MG	F	MG	MG
11	Compliance with regulations ($\mathscr{F}11$)	MG	F	MG	F	MG	MG	F	MG	MG	G
12	Resource utilization ($\mathscr{F}12$)	MG	MG	MG	F	MG	G	MG	MG	G	MG
13	Social responsibility ($\mathscr{F}13$)	MG	F	MG	F	MG	F	MG	F	MG	G
14	Technology adoption ($\mathscr{F}14$)	MG	G	G	MG	MG	G	G	MG	G	MG
	Criteria weight	H	VH	H	H	H					

Step 5: Compute the aggregations for factor ratings and criteria weights using the following equations:

$$\text{Factor ratings aggregation} = \mathring{FA}_{ij} = (\mathring{FA}^{1}_{ij}, \mathring{FA}^{x}_{ij}, \mathring{FA}^{y}_{ij}, \mathring{FA}^{Z}_{ij}) \tag{4.1}$$

where

$$\mathring{FA}^{1}_{ij} = \mathrm{Min}(\mathring{FA}^{1}_{ije}); \mathring{FA}^{x}_{ij} = \frac{1}{e}\sum\nolimits_{e=1}^{E}(\mathring{FA}^{x}_{ije});$$

$$\mathring{FA}^{y}_{ij} = \frac{1}{e}\sum\nolimits_{e=1}^{E}(\mathring{FA}^{y}_{ije}); \mathring{FA}^{z}_{ij} = \mathrm{Max}_{e}(\mathring{FA}^{z}_{ije})$$

The gathered inputs in the form of linguistic terms were further transformed into fuzzy numbers for investigational purposes. The transformed fuzzy numbers could then be aggregated by using Eq. (4.1).

Sample computation:

The gathered inputs for factor "$\mathscr{F}1$" under criteria "I" in linguistic terms are "G and MG." The inputs were transformed into the fuzzy numbers (7, 8, 8, 9) and (5, 6, 7, 8).

Then aggregation of "$\mathscr{F}1$" using Eq. (4.1) is as follows:

$$\mathring{FA}^{1}_{11} = \mathrm{Min}(7,5) = 5; \mathring{FA}^{2}_{11} = \frac{1}{2}(Sum(8,6)) = 7;$$

$$\mathring{FA}^{3}_{11} = \frac{1}{2}(Sum(8,7)) = 7.5 \text{ and } \mathring{FA}^{4}_{11} = \mathrm{Max}(9,8) = 9.$$

Because the weights were collected from the expert panel as a consensus opinion, there is no need for aggregation of weight.

Hence all individual factor ratings were aggregated by using Eq. (4.1) and are depicted in Table 4.3.

Table 4.3 represents the aggregated factor ratings in fuzzy numbers for all factors individually. Because the inputs for criteria weights are consensus opinion, there is no need for aggregation of the criteria weights. Hence the criteria weights were transformed into fuzzy numbers, and the weights are as follows: I—(0.7, 0.8, 0.8, 0.9), TR—(0.8, 0.9, 0.1, 0.1), F—(0.7, 0.8, 0.8, 0.9), C—(0.7, 0.8, 0.8, 0.9), and RoI—(0.7, 0.8, 0.8, 0.9).

Step 6: Computation of normalized decision matrix

The decision matrix needs to be normalized to avoid complexity in decision-making. The identified five criteria are advantageous to the industry practitioners. Hence all identified criteria were considered beneficial.

Hence the normalized decision matrix $\mathbb{N} = [\mathbb{N}_{ij}]_{m \times n}$ (Anand & Vinodh, 2018) is

$$[\mathbb{N}_{ij}] = (\mathbb{N}^{1}_{ij}, \mathbb{N}^{x}_{ij}, \mathbb{N}^{y}_{ij}, \mathbb{N}^{z}_{ij}) = (\mathring{FA}^{1}_{ij}/\mathring{FA}^{z+}_{j}, \mathring{FA}^{x}_{ij}/\mathring{FA}^{z+}_{j}, \mathring{FA}^{y}_{ij}/\mathring{FA}^{z+}_{j}, \mathring{FA}^{z}_{ij}/\mathring{FA}^{z+}_{j}) \tag{4.2}$$

TABLE 4.3 Aggregated fuzzy ratings for individual factors.

	I	TR	F	C	RoI
$\mathcal{F}1$	(5, 7, 7.5, 9)	(5, 7, 7.5, 9)	(7, 8.5, 9, 10)	(5, 7, 7.5, 9)	(5, 7, 7.5, 9)
$\mathcal{F}2$	(4, 5.5, 6, 8)	(5, 7, 7.5, 9)	(4, 5.5, 6, 8)	(4, 5, 5, 6)	(4, 6.5, 6.5, 9)
$\mathcal{F}3$	(7, 8.5, 9, 10)	(5, 7, 7.5, 9)	(4, 6.5, 6.5, 9)	(4, 5.5, 6, 8)	(5, 7, 7.5, 9)
$\mathcal{F}4$	(5, 7, 7.5, 9)	(7, 8.5, 9, 10)	(7, 8.5, 9, 10)	(5, 7, 7.5, 9)	(5, 7, 7.5, 9)
$\mathcal{F}5$	(5, 7, 7.5, 9)	(5, 7, 7.5, 9)	(7, 8, 8, 9)	(5, 7, 7.5, 9)	(7, 8, 8, 9)
$\mathcal{F}6$	(5, 6, 7, 8)	(7, 8, 8, 9)	(5, 7, 7.5, 9)	(5, 7, 7.5, 9)	(7, 8, 8, 9)
$\mathcal{F}7$	(5, 7, 7.5, 9)	(5, 7, 7.5, 9)	(7, 8, 8, 9)	(5, 7, 7.5, 9)	(5, 6, 7, 8)
$\mathcal{F}8$	(5, 7, 7.5, 9)	(5, 7, 7.5, 9)	(5, 7, 7.5, 9)	(4, 5.5, 6, 8)	(5, 7, 7.5, 9)
$\mathcal{F}9$	(5, 7, 7.5, 9)	(5, 7, 7.5, 9)	(4, 5.5, 6, 8)	(4, 5.5, 6, 8)	(5, 7, 7.5, 9)
$\mathcal{F}10$	(5, 7, 7.5, 9)	(5, 7, 7.5, 9)	(4, 5.5, 6, 8)	(4, 5.5, 6, 8)	(5, 6, 7, 8)
$\mathcal{F}11$	(4, 5.5, 6, 8)	(4, 5.5, 6, 8)	(5, 6, 7, 8)	(4, 5.5, 6, 8)	(5, 7, 7.5, 9)
$\mathcal{F}12$	(5, 6, 7, 8)	(4, 5.5, 6, 8)	(5, 7, 7.5, 9)	(5, 6, 7, 8)	(5, 7, 7.5, 9)
$\mathcal{F}13$	(4, 5.5, 6, 8)	(4, 5.5, 6, 8)	(4, 5.5, 6, 8)	(4, 5.5, 6, 8)	(5, 7, 7.5, 9)
$\mathcal{F}14$	(5, 7, 7.5, 9)	(5, 7, 7.5, 9)	(5, 7, 7.5, 9)	(5, 7, 7.5, 9)	(5, 7, 7.5, 9)

where

$$\mathring{FA}_j^{z+} = \mathrm{Max}_e(\mathring{FA}_{ij}^{z}); \mathring{FA}_j^{1-} = \mathrm{Min}_e(\mathring{FA}_{ij}^{1})$$

The normalized decision matrix was computed by using Eq. (4.2). Following is a sample calculation: For factor "$\mathcal{F}1$" under criteria "I"

$$\mathring{FA}_j^{4+} = \mathrm{Max}_i(\mathring{FA}_{ij}^{z}) = \mathrm{Max}(9, 8, 10, 9, 9, 8, 9, 9, 9, 9, 8, 8, 8, 9) = 10$$

$$\text{Normalization } [N]_{1\times 1} = \left(\frac{5}{10}, \frac{7}{10}, \frac{7.5}{10}, \frac{9}{10}\right) = (0.5, 0.7, 0.75, 0.9)$$

Similarly, the entire aggregated matrix was normalized for every criterion using Eq. (4.2). The normalized matrix is depicted in Table 4.4.

Step 7: Calculation of weighted normalized decision matrix

After the normalized decision matrix has been attained, each normalized factor rating in the matrix was multiplied with each corresponding criteria weight.

The weighted normalized matrix is as follows:

$$N\omega = [N\omega_{ij}] = [\omega_j] * [N_{ij}]_{m\times n} \quad (4.3)$$

TABLE 4.4 Normalized decision matrix.

	I	TR	F	C	RoI
$\mathscr{F}1$	(0.5, 0.7, 0.75, 0.9)	(0.5, 0.7, 0.75, 0.9)	(0.7, 0.85, 0.9, 1)	(0.56, 0.78, 0.83, 1)	(0.56, 0.78, 0.83, 1)
$\mathscr{F}2$	(0.4, 0.55, 0.6, 0.8)	(0.5, 0.7, 0.75, 0.9)	(0.4, 0.55, 0.6, 0.8)	(0.44, 0.56, 0.56, 0.67)	(0.44, 0.72, 0.72, 1)
$\mathscr{F}3$	(0.7, 0.85, 0.9, 1)	(0.5, 0.7, 0.75, 0.9)	(0.4, 0.65, 0.65, 0.9)	(0.44, 0.61, 0.67, 0.89)	(0.56, 0.78, 0.83, 1)
$\mathscr{F}4$	(0.5, 0.7, 0.75, 0.9)	(0.7, 0.85, 0.9, 1)	(0.7, 0.85, 0.9, 1)	(0.56, 0.78, 0.83, 1)	(0.56, 0.78, 0.83, 1)
$\mathscr{F}5$	(0.5, 0.7, 0.75, 0.9)	(0.5, 0.7, 0.75, 0.9)	(0.7, 0.8, 0.8, 0.9)	(0.56, 0.78, 0.83, 1)	(0.78, 0.89, 0.89, 1)
$\mathscr{F}6$	(0.5, 0.6, 0.7, 0.8)	(0.7, 0.8, 0.8, 0.9)	(0.5, 0.7, 0.75, 0.9)	(0.56, 0.78, 0.83, 1)	(0.78, 0.89, 0.89, 1)
$\mathscr{F}7$	(0.5, 0.7, 0.75, 0.9)	(0.5, 0.7, 0.75, 0.9)	(0.7, 0.8, 0.8, 0.9)	(0.56, 0.78, 0.83, 1)	(0.56, 0.67, 0.78, 0.89)
$\mathscr{F}8$	(0.5, 0.7, 0.75, 0.9)	(0.5, 0.7, 0.75, 0.9)	(0.5, 0.7, 0.75, 0.9)	(0.44, 0.61, 0.67, 0.89)	(0.56, 0.78, 0.83, 1)
$\mathscr{F}9$	(0.5, 0.7, 0.75, 0.9)	(0.5, 0.7, 0.75, 0.9)	(0.4, 0.55, 0.6, 0.8)	(0.44, 0.61, 0.67, 0.89)	(0.56, 0.78, 0.83, 1)
$\mathscr{F}10$	(0.5, 0.7, 0.75, 0.9)	(0.5, 0.7, 0.75, 0.9)	(0.4, 0.55, 0.6, 0.8)	(0.44, 0.61, 0.67, 0.89)	(0.56, 0.67, 0.78, 0.89)
$\mathscr{F}11$	(0.4, 0.55, 0.6, 0.8)	(0.4, 0.55, 0.6, 0.8)	(0.5, 0.6, 0.7, 0.8)	(0.44, 0.61, 0.67, 0.89)	(0.56, 0.78, 0.83, 1)
$\mathscr{F}12$	(0.5, 0.6, 0.7, 0.8)	(0.4, 0.55, 0.6, 0.8)	(0.5, 0.7, 0.75, 0.9)	(0.56, 0.67, 0.78, 0.89)	(0.56, 0.78, 0.83, 1)
$\mathscr{F}13$	(0.4, 0.55, 0.6, 0.8)	(0.4, 0.55, 0.6, 0.8)	(0.4, 0.55, 0.6, 0.8)	(0.44, 0.61, 0.67, 0.89)	(0.56, 0.78, 0.83, 1)
$\mathscr{F}14$	(0.5, 0.7, 0.75, 0.9)	(0.5, 0.7, 0.75, 0.9)	(0.5, 0.7, 0.75, 0.9)	(0.56, 0.78, 0.83, 1)	(0.56, 0.78, 0.83, 1)

A sample calculation of weighted normalization for factor "$\mathscr{F}1$" corresponding to criteria "I" is

$$[N\omega_{11}] = (0.5, 0.7, 0.75, 0.9) * (0.7, 0.8, 0.8, 0.9) = (0.35, 0.56, 0.6, 0.81)$$

Similarly, the weighted normalized matrix was calculated for each criterion by using Eq. (4.3). The weighted normalized values are listed in Table 4.5.

TABLE 4.5 Weighted normalized decision matrix.

	I	TR	F	C	RoI
$\mathscr{F}1$	(0.35, 0.56, 0.6, 0.81)	(0.4, 0.63, 0.08, 0.09)	(0.49, 0.68, 0.72, 0.9)	(0.39, 0.62, 0.66, 0.9)	(0.39, 0.62, 0.66, 0.9)
$\mathscr{F}2$	(0.28, 0.44, 0.48, 0.72)	(0.4, 0.63, 0.08, 0.09)	(0.28, 0.44, 0.48, 0.72)	(0.31, 0.45, 0.45, 0.6)	(0.31, 0.58, 0.58, 0.9)
$\mathscr{F}3$	(0.49, 0.68, 0.72, 0.9)	(0.4, 0.63, 0.08, 0.09)	(0.28, 0.52, 0.52, 0.81)	(0.31, 0.49, 0.54, 0.8)	(0.39, 0.62, 0.66, 0.9)
$\mathscr{F}4$	(0.35, 0.56, 0.6, 0.81)	(0.56, 0.77, 0.09, 0.1)	(0.49, 0.68, 0.72, 0.9)	(0.39, 0.62, 0.66, 0.9)	(0.39, 0.62, 0.66, 0.9)
$\mathscr{F}5$	(0.35, 0.56, 0.6, 0.81)	(0.4, 0.63, 0.08, 0.09)	(0.49, 0.64, 0.64, 0.81)	(0.39, 0.62, 0.66, 0.9)	(0.55, 0.71, 0.71, 0.9)
$\mathscr{F}6$	(0.35, 0.48, 0.56, 0.72)	(0.56, 0.72, 0.08, 0.09)	(0.35, 0.56, 0.6, 0.81)	(0.39, 0.62, 0.66, 0.9)	(0.55, 0.71, 0.71, 0.9)
$\mathscr{F}7$	(0.35, 0.56, 0.6, 0.81)	(0.4, 0.63, 0.08, 0.09)	(0.49, 0.64, 0.64, 0.81)	(0.39, 0.62, 0.66, 0.9)	(0.39, 0.54, 0.62, 0.8)
$\mathscr{F}8$	(0.35, 0.56, 0.6, 0.81)	(0.4, 0.63, 0.08, 0.09)	(0.35, 0.56, 0.6, 0.81)	(0.31, 0.49, 0.54, 0.8)	(0.39, 0.62, 0.66, 0.9)
$\mathscr{F}9$	(0.35, 0.56, 0.6, 0.81)	(0.4, 0.63, 0.08, 0.09)	(0.28, 0.44, 0.48, 0.72)	(0.31, 0.49, 0.54, 0.8)	(0.39, 0.62, 0.66, 0.9)
$\mathscr{F}10$	(0.35, 0.56, 0.6, 0.81)	(0.4, 0.63, 0.08, 0.09)	(0.28, 0.44, 0.48, 0.72)	(0.31, 0.49, 0.54, 0.8)	(0.39, 0.54, 0.62, 0.8)
$\mathscr{F}11$	(0.28, 0.44, 0.48, 0.72)	(0.32, 0.5, 0.06, 0.08)	(0.35, 0.48, 0.56, 0.72)	(0.31, 0.49, 0.54, 0.8)	(0.39, 0.62, 0.66, 0.9)
$\mathscr{F}12$	(0.35, 0.48, 0.56, 0.72)	(0.32, 0.5, 0.06, 0.08)	(0.35, 0.56, 0.6, 0.81)	(0.39, 0.54, 0.62, 0.8)	(0.39, 0.62, 0.66, 0.9)
$\mathscr{F}13$	(0.28, 0.44, 0.48, 0.72)	(0.32, 0.5, 0.06, 0.08)	(0.28, 0.44, 0.48, 0.72)	(0.31, 0.49, 0.54, 0.8)	(0.39, 0.62, 0.66, 0.9)
$\mathscr{F}14$	(0.35, 0.56, 0.6,0.81)	(0.4, 0.63, 0.08, 0.09)	(0.35, 0.56, 0.6, 0.81)	(0.39, 0.62, 0.66, 0.9)	(0.39, 0.62, 0.66, 0.9)

Step 7: Computation of ideal positive and nadir solutions of every criterion

Here, all criteria were identified as advantageous. Hence the fuzzy deal positive solution is as follows:

$$I^{+} = \mathrm{Max}(\mathbb{N}\omega_{ij}) \tag{4.4}$$

The nadir solution is as follows:

$$I^{-} = \mathrm{Min}(\mathbb{N}\omega_{ij}) \tag{4.5}$$

TABLE 4.6 Ideal positive and nadir solutions of each criterion.

	I	TR	F	C	RoI
Ideal positive solution	(0.49, 0.68, 0.72, 0.9)	(0.56, 0.77, 0.09, 0.1)	(0.49, 0.68, 0.72, 0.9)	(0.39, 0.62, 0.66, 0.9)	(0.55, 0.71, 0.71, 0.9)
Nadir solution	(0.28, 0.44, 0.48, 0.72)	(0.32, 0.5, 0.06, 0.08)	(0.28, 0.44, 0.48, 0.72)	(0.31, 0.45, 0.45, 0.6)	(0.31, 0.54, 0.58, 0.8)

The ideal positive and nadir solutions were computed by using Eqs. (4.4) and (4.5) for every criterion and are listed in Table 4.6.

Step 8: Computation of separation measurement from the fuzzy ideal positive solution and nadir solution corresponding to every criterion (Anand & Vinodh, 2018).

Every factor separation measurement from the ideal positive solution is as follows:

$$Sm_i^+ = \sqrt{\sum_{i=1}^{n} \frac{1}{4}\left(\mathbb{N}\omega_{ij} - I^+\right)^2} \tag{4.6}$$

Every factor separation measurement the from nadir solution is as follows:

$$Sm_i^- = \sqrt{\sum_{i=1}^{n} \frac{1}{4}\left(\mathbb{N}\omega_{ij} - I^-\right)^2} \tag{4.7}$$

The separation measurement was computed for all individual factors by using Eqs. (4.6) and (4.7). The attained measurements are listed in Table 4.7.

Step 9: Calculation of closeness coefficient of every factor

$$\mathbb{CC}_{\mathrm{i}} = \frac{sm_i^-}{sm_i^+ + sm_i^-} \tag{4.8}$$

The closeness coefficient for every factor was computed by using Eq. (4.8). The attained coefficients of all factors are listed in Table 4.8.

Step 10: Finally, ranking among factors influencing the adoption of SSM systems needs to be attained. The prioritization of the factors influencing SSM systems adoption is seen in Table 4.8.

Table 4.8 shows the closeness coefficient of factors and ranking among the factors. The derived ranking is as follows: integration < reliability < automation < top management commitment < increased productivity < technology adoption < economy < 3R and 6R adoption < resource utilization < energy consumption < workforce training < compliance to regulations < social responsibility < security.

TABLE 4.7 Separation measurements from ideal positive and nadir solutions.

	From ideal positive solution					From nadir solution					Separation measurement	
	I	TR	F	C	RoI	I	TR	F	C	RoI	di +	di −
$\mathcal{F}1$	0.1188	0.1065	0	0	0.0951	0.1022	0.0771	0.2189	0.2058	0.0854	0.3205	0.6895
$\mathcal{F}2$	0.2189	0.1065	0.2189	0.2058	0.1512	0	0.0771	0	0	0.0539	0.9013	0.131
$\mathcal{F}3$	0	0.1065	0.1716	0.1092	0.0951	0.2189	0.0771	0.0634	0.1115	0.0854	0.4825	0.5564
$\mathcal{F}4$	0.1188	0	0	0	0.0951	0.1022	0.1815	0.2189	0.2058	0.0854	0.214	0.7939
$\mathcal{F}5$	0.1188	0.1065	0.0634	0	0	0.1022	0.0771	0.1716	0.2058	0.1684	0.2888	0.7251
$\mathcal{F}6$	0.1715	0.026	0.1188	0	0	0.0568	0.1632	0.1022	0.2058	0.1684	0.3163	0.6964
$\mathcal{F}7$	0.1188	0.1065	0.0634	0	0.1347	0.1022	0.0771	0.1716	0.2058	0.0447	0.4235	0.6015
$\mathcal{F}8$	0.1188	0.1065	0.1188	0.1092	0.0951	0.1022	0.0771	0.1022	0.1115	0.0854	0.5486	0.4785
$\mathcal{F}9$	0.1188	0.1065	0.2189	0.1092	0.0951	0.1022	0.0771	0	0.1115	0.0854	0.6486	0.3763
$\mathcal{F}10$	0.1188	0.1065	0.2189	0.1092	0.1347	0.1022	0.0771	0	0.1115	0.0447	0.6882	0.3356
$\mathcal{F}11$	0.2189	0.1815	0.1715	0.1092	0.0951	0	0	0.0568	0.1115	0.0854	0.7762	0.2537
$\mathcal{F}12$	0.1715	0.1815	0.1188	0.0671	0.0951	0.0568	0	0.1022	0.1444	0.0854	0.634	0.3888
$\mathcal{F}13$	0.2189	0.1815	0.2189	0.1092	0.0951	0	0	0	0.1115	0.0854	0.8237	0.1969
$\mathcal{F}14$	0.1188	0.1065	0.1188	0	0.0951	0.1022	0.0771	0.1022	0.2058	0.0854	0.4394	0.5728

TABLE 4.8 Closeness coefficient and ranking of factors.

Factors	**Top management commitment**	**Security**	**Economy**	**Integration**	**Reliability**	**Automation**	**Increased productivity**
Closeness coefficient	0.6827	0.1269	0.5356	0.7877	0.7152	0.6877	0.5868
Rank	4	14	7	1	2	3	5
Factors	**3R and 6R adoption**	**Energy consumption**	**Workforce training**	**Compliance with regulations**	**Resource utilization**	**Social responsibility**	**Technology adoption**
Closeness coefficient	0.4659	0.3672	0.3278	0.2463	0.3801	0.1929	0.5659
Rank	8	10	11	12	9	13	6

4.4 Conclusion

Industries are facing global pressure resulting from the occurrence of dynamic changes. Governments have introduced many efforts to assist industries in preparing for the impending globalization. The most significant new phase changes at the base of manufacturing industries are Industry 4.0 and SM. The study described in this chapter can assist industry people in understanding the current reality of Industry 4.0, SM, and sustainable manufacturing. Several SSM techniques may be used at the industrial level for enhanced revenues. The SSM system is more significant to enhance production capacity to fulfill customer demand. To attain this, the industries need to adopt SSM systems. Before implementing SSM, industry practitioners need to know the most significant factors influencing SSM. Hence we identified and examined the factors influencing SSM using a fuzzy TOPSIS approach. Based on the results attained in the investigation process, integration, reliability, and automation are the top factors that influence the adoption of SSM systems in the industry. This investigation will facilitate industry practitioners in identifying the most influencing factors of SSM. Hence industry practitioners can focus on these factors during the adoption of SSM systems.

References

Abubakr, M., Abbas, A. T., Tomaz, I., Soliman, M. S., Luqman, M., & Hegab, H. (2020). Sustainable and smart manufacturing: An integrated approach. *Sustainability (Switzerland), 12*(6), 1–19. Available from https://doi.org/10.3390/su12062280.

Aggarwal, A., Gupta, S., Jamwal, A., Agrawal, R., Sharma, M., & Dangayach, G. S. (2021). Adoption of smart and sustainable manufacturing practices: An exploratory study of Indian manufacturing companies. *Proceedings of the Institution of Mechanical Engineers, Part B: Journal of Engineering Manufacture, 236*(5), 586–602. Available from https://doi.org/10.1177/09544054211040646.

Agrawal, R., Wankhede, V. A., Kumar, A., Upadhyay, A., & Garza-Reyes, J. A. (2021). Nexus of circular economy and sustainable business performance in the era of digitalization. *International Journal of Productivity and Performance Management, 71*(3), 748–774. Available from https://doi.org/10.1108/IJPPM-12-2020-0676.

Anand, M. B., & Vinodh, S. (2018). Application of fuzzy AHP – TOPSIS for ranking additive manufacturing processes for microfabrication. *Rapid Prototyping Journal, 24*(2), 424–435. Available from https://doi.org/10.1108/RPJ-10-2016-0160.

Barrientos, S., & Gereffi, G. (2011). A new paradigm for a changing world. *International Labour Review, 150*(3), 319–340.

Chavarría-Barrientos, D., Batres, R., Wright, P. K., & Molina, A. (2018). A methodology to create a sensing, smart and sustainable manufacturing enterprise. *International Journal of Production Research, 56*(1–2), 584–603. Available from https://doi.org/10.1080/00207543.2017.1386333.

Dib, M. A. D. S., Ribeiro, B., & Prates, P. (2021). Federated learning as a privacy-providing machine learning for defect predictions in smart manufacturing. *Smart and Sustainable Manufacturing Systems, 5*(1), 1–17. Available from https://doi.org/10.1520/SSMS20200029.

Edwards, C. (2021). Real-time advanced analytics, automated production systems, and smart industrial value creation in sustainable manufacturing internet of things. *Journal of Self-Governance and Management Economics*, *9*(2), 32–41.

Guo, Y., Wang, N., Xu, Z. Y., & Wu, K. (2020). The internet of things-based decision support system for information processing in intelligent manufacturing using data mining technology. *Mechanical Systems and Signal Processing*, *142*, 106630. Available from https://doi.org/10.1016/J.YMSSP.2020.106630.

Isharyani, M. E., Wijaya, Y., Sitania, F. D., Fahrul Hassan, M., Rahman, M., Arifin, A., ... Okopi, M. (2021). Urbanization and sustainable growth of Urban Kano, Nigeria. *IOP Conference Series: Earth and Environmental Science*, *665*(1), 012063. Available from https://doi.org/10.1088/1755-1315/665/1/012063.

Lenz, J., MacDonald, E., Harik, R., & Wuest, T. (2020). Optimizing smart manufacturing systems by extending the smart products paradigm to the beginning of life. *Journal of Manufacturing Systems*, *57*, 274–286. Available from https://doi.org/10.1016/J.JMSY.2020.10.001.

Lepore, D., Micozzi, A., & Spigarelli, F. (2021). Industry 4.0 accelerating sustainable manufacturing in the COVID-19 era: Assessing the readiness and responsiveness of italian regions. *Sustainability*, *13*(5), 2670. Available from https://doi.org/10.3390/SU13052670, 2021, Vol. 13, Page 2670.

Malaga, A., & Vinodh, S. (2021a). Evaluation of smart manufacturing performance using a grey theory-based approach: A case study. *Grey Systems: Theory and Application*. Available from https://doi.org/10.1108/gs-04-2021-0049, ahead-of-p(ahead-of-print).

Malaga, A., & Vinodh, S. (2021b). Benchmarking smart manufacturing drivers using Grey TOPSIS and COPRAS-G approaches. *Benchmarking*, *28*(10), 2916–2951. Available from https://doi.org/10.1108/BIJ-12-2020-0620/FULL/PDF.

Meng, Y., Yang, Y., Chung, H., Lee, P. H., & Shao, C. (2018). Enhancing sustainability and energy efficiency in smart factories: A review. *Sustainability*, *10*(12), 4779. Available from https://doi.org/10.3390/SU10124779, 2018, Vol. 10, Page 4779.

Molina, A., Ponce, P., Miranda, J., & Cortés, D. (2021). *Sensing, smart and sustainable manufacturing processes. Enabling Systems for Intelligent Manufacturing in Industry 4.0* (pp. 141–226). Cham: Springer. Available from https://doi.org/10.1007/978-3-030-65547-1_4.

Raman, A. S., Haapala, K. R., Raoufi, K., Linke, B. S., Bernstein, W. Z., & Morris, K. C. (2020). Defining near-term to long-term research opportunities to advance metrics, models, and methods for smart and sustainable manufacturing. *Smart and Sustainable Manufacturing Systems*, *4*(2), 24. Available from https://doi.org/10.1520/SSMS20190047.

Shukla, O. J., Jangid, V., Siddh, M. M., Kumar, R., & Soni, G. (2017). Evaluating key factors of sustainable manufacturing in Indian automobile industries using analytic hierarchy process (AHP). 2017 International Conference on Advances in Mechanical, Industrial, Automation and Management Systems, AMIAMS 2017 – Proceedings, 42–47. Available from https://doi.org/10.1109/AMIAMS.2017.8069186.

Singh, S., Khamba, J. S., & Singh, D. (2021). Role and scope of overall equipment effectiveness implementation in Indian sugarmill industries: A justified approach. *Proceedings of the Institution of Mechanical Engineers, Part E: Journal of Process Mechanical Engineering*, *236*(2), 546–555. Available from https://doi.org/10.1177/09544089211050173.

Sunmola, F. T., & Apeji, U. D. (2020). Blockchain characteristics for sustainable supply chain visibility. Proceedings of the 5th NA International Conference on Industrial Engineering and Operations Management Detroit, Michigan, USA, August 10–14.

Susana Ferreiro, S. F., Larreina, J., Tena, M., Leunda, J., Garmendia, I., & Arnaiz, A. (2021). Artificial intelligence methodology for smart and sustainable manufacturing industry. *IFAC-PapersOnLine*, *54*(1), 1041–1046. Available from https://doi.org/10.1016/j.ifacol.2021.08.201.

Suvarna, M., Büth, L., Hejny, J., Mennenga, M., Li, J., Ng, Y. T., … Wang, X. (2020). Smart manufacturing for smart cities—overview, insights, and future directions. *Advanced Intelligent Systems*, *2*(10), 2000043. Available from https://doi.org/10.1002/AISY.202000043.

Vinodh, S., Thiagarajan, A., & Mulanjur, G. (2014). Lean concept selection using modified fuzzy TOPSIS: A case study. *International Journal of Services and Operations Management*, *18*(3), 342–357. Available from https://doi.org/10.1504/IJSOM.2014.062235.

Wang, B., Hu, S. J., Sun, L., & Freiheit, T. (2020). Intelligent welding system technologies: State-of-the-art review and perspectives. *Journal of Manufacturing Systems*, *56*, 373–391. Available from https://doi.org/10.1016/J.JMSY.2020.06.020.

Wu, B., Gao, B., Xu, W., Wang, H., Yi, Y., & R, P. (2022). Sustainable food smart manufacturing technology. *Information Processing & Management*, *59*(1), 102754. Available from https://doi.org/10.1016/J.IPM.2021.102754.

Yang, B., Usman, M., & Jahanger, A. (2021). Do industrialization, economic growth and globalization processes influence the ecological footprint and healthcare expenditures? Fresh insights based on the STIRPAT model for countries with the highest healthcare expenditures. *Sustainable Production and Consumption*, *28*, 893–910. Available from https://doi.org/10.1016/J.SPC.2021.07.020.

CHAPTER

5

Catching up with sustainable development in emerging markets through financial innovation

Mirela Panait[1,2], *Eglantina Hysa*[3] *and Lukman Raimi*[4]

[1]Petroleum-Gas University of Ploieşti, Ploieşti, Romania, [2]Institute of National Economy, Bucharest, Romania, [3]Department of Economics, Epoka University, Tirana, Albania, [4]Universiti Brunei Darussalam (UBD), Brunei

5.1 Introduction

Economic activity is an important area in terms of innovation and sustainable development, and the instruments and mechanisms of the financial market are gaining new value for financing these particularly complex processes and with many challenges for stakeholders (Stancu & Iacovoiu, 2015; Iacovoiu & Stancu, 2017). Globally, the financial market is crucial to the smooth operation of capitalist economies, as it provides a regulated platform for trading all types of assets or securities, such as stocks, bonds, foreign currency, and derivatives (Johnson et al., 2003; Duffie, 2014). Financial market comprise banking, insurance, and capital markets. The capital market segment provides a platform for buyers and sellers (as individuals and institutions) to trade financial securities such as bonds and stocks (The Economic Times, 2021). The money market, by contrast, is the platform for trading in financial instruments, essentially commercial papers with high liquidity and short-term maturities

Innovation and Sustainable Manufacturing
DOI: https://doi.org/10.1016/B978-0-12-819513-0.00003-2

(The Economic Times, 2021). The globalization and emergence of digital technologies are disrupting all facets of the global economy, including financial market operations and instruments (Brezeanu et al., 2005; Stancu et al., 2017). Consequently, the financial market is one of the most attractive and constantly changing areas of economic life, continually pursuing the improvement and accessibility of instruments offered to investors, adapting these products to the needs of users through the emergence of new financial products, simplifying trading activities, reducing trading costs, and so on.

The stock market innovation process is characteristic of the largest markets, determining the appearance of modern financial products, such as derivatives for weather indices; derivatives for the real estate sector; environment, social, and governance (ESG products); and sustainability indices. The stock market innovation process in Europe is supported by the most important stock exchanges, such as Euronext, London Stock Exchange, and Deutsche Borse (Matei, 2013).

In the process of launching and developing financial products, stock market specialists need to keep in mind that the long-term evolution of commodity and stock exchanges is based largely on the activities of institutional investors. In addition to interdomain competition, there is interregional competition between the US and European stock exchanges. The sometimes hostile competition between the stock exchanges on the two continents has also generated friendly mergers, the most notable example being the merger between the New York Stock Exchange (NYSE) and Euronext. The result was the establishment of the world's largest stock exchange group, namely, NYSE Euronext, which offers spot and derivative products for European and US markets.

Financial innovation has been fueled in recent decades by the need to launch sustainable development instruments. Sustainable development is a very complex process with many economic, social, and technical challenges that require considerable financial resources that must be channeled to certain sectors in order to reduce their negative impact on the environment. The interest of companies that generate negative externalities is growing for financial instruments and products that promote sustainability. In addition, capital market players, that is, retail investors, institutional investors, and stock exchanges, have become increasingly attentive to the nonfinancial performance of listed companies. Therefore sustainable development and corporate social responsibility are the watchwords in the business world in general and in the capital market in particular (Panait et al., 2014; Brezoi, 2018; Ionescu, 2019; Gigauri, 2021; Gigauri & Djakeli, 2021; Khan et al., 2021; Răzvan, 2021; Vasile et al., 2021; Voica & Stancu 2021).

5.2 Literature review and theoretical framework

Financial market innovation requires institutional, legal, and regulatory frameworks. This fact has been explicated by a number of scholars. According to Alfaro et al. (2019), role institutions cannot be underplayed because "on one side the institutions (sets of rules), organizational culture, governmental policies, financial markets and external business environment are essential for the establishment of appropriate conditions for innovation, and on the other side, the entrepreneur plays an important role in the creation of new technology-based firms and technological innovation with the help of this accumulated knowledge (technological, managerial, risk management, financial, etc.) and culture."

Furthermore, Çankaya (2014) opined that financial innovation is influenced by macroeconomic conditions, and the latter has an impact on the macroeconomic environment. Financial innovation is a technology-driven dynamism that should address customers' needs and be supported by appropriate policies and regulations. As such, it contributes to increased competition and pushes the idea in a broader aspect that has global impacts. In the conceptual dimension this study is focused on the theory of innovation that is specific to financial innovation. Therefore we try to give a general prescript based on the literature by focusing on the definitions, types, and functions of financial innovations analyzed from the financial system perspective. Then we try to develop a proposed framework that links financial innovation with sustainable development: - the economic-environmental-social triangle.

Among many types of innovation we find that financial innovation is one of the special types that has been widely discussed in theoretical and empirical studies. In the early 1930s, Schumpeter defined innovation as the introduction of a new or qualitative change in existing products, processes, markets, sources of supply, and organizations. Further studies have classified innovation on the basis of different approaches and disciplines. For example, Baregheh et al. (2009) propose the key attributes of innovations shown in Fig. 5.1.

Even though financial innovations are not a new issue (Michalopoulos et al., 2009), financial innovation has given a new trend to modern financial systems, and its importance has been widely recognized (Ajide, 2016). However, what are the types of financial innovation? Many studies have considered financial innovation only in the narrow aspect of product innovation. Other studies, such as those of Merton (1995), Llewellyn (2009), and Lerner and Tufano (2011), have discussed upon a broader aspect the variety of financial innovations.

Financial system innovations or financial institutional innovations are associated with the financial sector as a whole. They are related to changes in business structures up to the establishment and design of

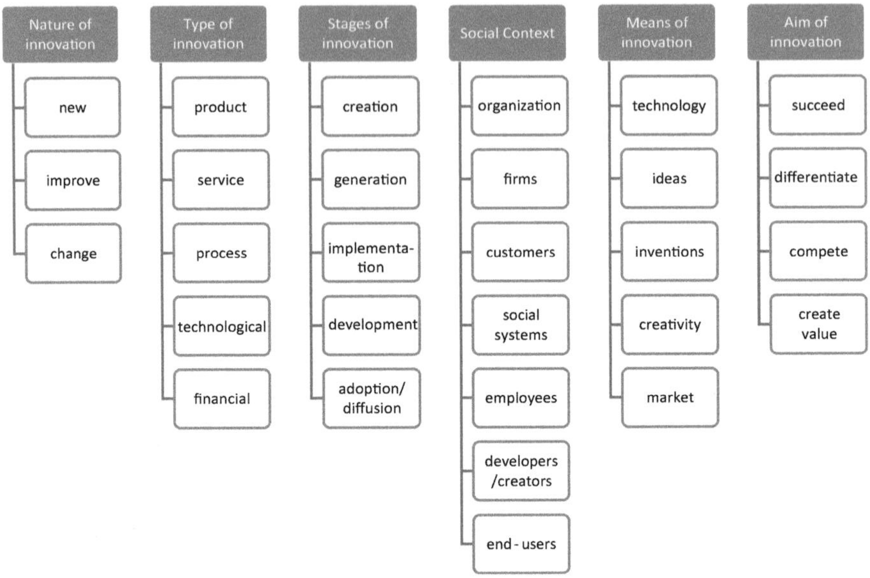

FIGURE 5.1 Key attributions of innovation. Source: *Adapted from Baregheh, A., Rowley, J., & Sambrook, S. (2009). Towards a multidisciplinary definition of innovation.* Management Decision. *47(8) https://doi.org/10.1108/00251740910984578 and developed by the authors.*

new types of financial intermediaries that are connected, for instance, with changes in the legal and supervisory framework of financial systems. This is also supported by Ajide (2016), who stated that financial innovation should be linked with clear policies that would drive competition and efficiency in the banking industry; as such, they would ensure effective functioning of the financial system. Moreover, financial innovation is expected to enhance the financial system by maintaining its stability, increasing its efficiency in performing its core functions, and providing services and instruments that are better adjusted to the system participants' needs and goals (Błach, 2020).

Financial process innovations are involved with the introduction of new business processes that push and lead to an increase in efficiency, new business dimensions, market expansion, and so on. Product innovations include the introduction of specific tools in the financial system, such as new credit, deposits, leasing, insurance, and other financial products (Pyka, 2013; Frame & White, 2014). Product innovations are launched and introduced in the market to better respond to changes in financial systems and to support the enhancement of efficiency. All three types of innovation are tightly linked with technology-driven financial innovation, which provides a mechanism to finance innovative technological projects, while at the same time, technological progress creates a fundamental basis for new financial solutions (Pyka, 2013; Kapoor and Mention, 2015). As demonstrated in the left-hand side of Fig. 5.2, financial innovation is composed of all the above-mentioned upgrading or new innovations with regard to technology, systems/institutions, products and processes.

In the right-hand side of Fig. 5.2, we show that at the core of sustainable development lies financial innovation. The study of Hysa et al. (2020) highlighted the importance of innovation as an emergency and intermediate tool for sustainable development. Now, this study, as a "subconsequence" of innovation, comes up with a developed approach to the impact of innovation of the whole sustainable development structure. Based on other studies, such as Evstigneev et al. (2004) and Nesvetailova (2014), which state that the ability of the financial system to create innovations leads to transformations in the socioeconomic system and the political and legal system, we agree and propose that for sustainable development, financial innovation is a "must." This is supported by Michalopoulos et al. (2009), Lerner and Tufano (2011), Allen (2012), and Laeven et al. (2015), who stated that financial innovations are essential for technological innovations and economic growth, not only for individual entities but also for the entire society.

As financial innovation comprises the introduction and promotion of financial products and services, the development of new processes, and the interaction with customers and the development of new structures for financial institutions, innovation remains crucial in its role as a force for sustainable development and social inclusion (Mention, 2011; Salas, 2009). If properly distributed, financial innovation can reduce the gap of competitiveness and the uneven knowledge gap among countries. According to Nickell (1996), competition can be seen as a stimulus to increase downward pressure on the costs of services or products, reduce managerial laziness, and even improve the level of technology and

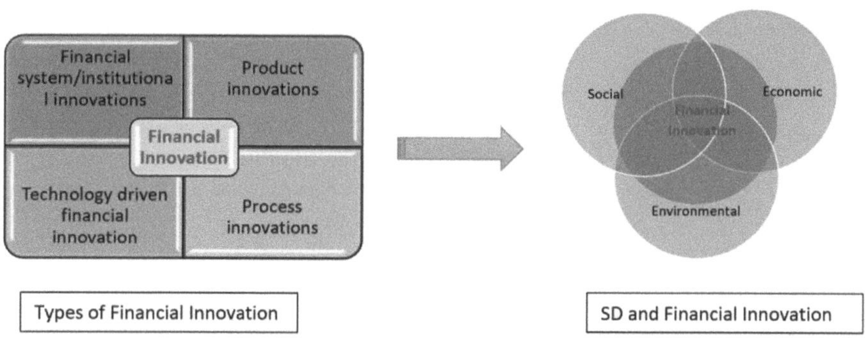

FIGURE 5.2 Conceptual framework of sustainable development through financial innovation. Source: *Developed by the authors.*

financial innovation. Therefore competition can push forward research and development with regard to technology, which supports the willingness to innovate constantly to be efficient and effective in the production of higher-quality products and to minimize costs to maintain or increase their market share and make more profits (Ajide & Ajileye, 2015; Motta, 2004).

As the number of end users and beneficiaries from the application of financial innovations is broad, for instance, financial institutions themselves, governments, large corporations, firms, and households, we expect that the effect of innovation is large-scaled and often has a spiral effect (multiplying effect). This approach is consistent with the rational efficiency hypothesis, "which states that financial innovations are created and implemented to generate profit for their users." Shiller (2013) addressed the importance of financial innovations to achieving society's goals. In their study, Beck et al. (2016) confirmed that there was a positive net effect of financial innovation on economic growth. This supports our proposed framework, which relates sustainable development to a good foundation of financial innovation, reinforcing the reduction of the existing imperfections in the financial system by lowering the risk level, reducing transaction costs, closing the information gap, and minimizing tax payments (Błach, 2020).

In general, we consider that financial innovations improve the firm's ability to use opportunities and to become more effective against threat protection. Thus sustainable financial innovations drive the enhancement of financial management and boost the firm to create value and become more interactive.

5.3 Methods and analysis

Given the paucity of literature on financial innovation in emerging markets, we purposely adopted qualitative research (an interpretivist paradigm) to provide deeper insight into the discussion on the nexus of sustainable development and financial innovation. We sourced and compiled different views, perspectives, and pieces of information from scholarly articles, working papers, and online resources, using the approach of Marszk and Lechman (2021). After the selection of scholarly materials, a content analysis was used to review, integrate, and synthesize the extracted textual information following the methodological and logical procedures for quantifying the contents of texts and writings (Saunders et al., 2016; Denscombe, 2017). For more clarity we followed a four-stage process:

1. Stage 1: Data sourcing. At this stage, we sourced the required numeric secondary data on financial markets, innovative financial products, and sustainable development to address our chapter

objectives. For data sourcing, we previewed explored Science Direct and Google Scholar, from which over 120 articles were downloaded, from which over 81 were selected based on the purposive sampling technique. The two selection criteria that we used are as follows:
   a. The selected articles discuss financial markets and sustainable development from the perspective of emerging markets.
   b. The selected articles provide insights into the financial market and digitalization theories.
2. Stage 2: Data development and conversion. At this stage, we compiled insightful pieces of information on financial markets and sustainable development based on relevance, recency, and suitability in readiness for analysis data.
3. Stage 3: Data analysis. Here, the insightful pieces of information that we compiled were critically reviewed, appraised, synthesized, and contextualized by using content analysis.
4. Stage 4: Contextualization using tables and figures. In the final stage, the valuable insights into digital financial markets and sustainable development were contextualized by using tables for easy comprehension by readers. This methodological approach is supported by Williams and Shepherd (2017) and Marszk and Lechman (2021).

5.4 Sustainable financing for sustainable developing

One of the financial innovations that was invented by the global finance ecosystem is the concept of sustainable financing. The term *sustainable finance* (SF) draws its meaning and dimensions and focuses on sustainability principles, such as the triple theme of sustainable development; environmental, social, and governance (ESG) dimensions; sustainable development goals (SDGs); and equator principles (Raimi et al., 2022). In definite terms, SF is the integration of aspects of sustainability in the decision-making processes of financial market actors, financial market policy, and related institutional and market arrangements that contribute to the achievement of strong, sustainable, balanced, and inclusive growth (Sommer, 2020). Dirk Schoenmaker (2017) provides a captivating description, defining sustainable finance as a unique financial option that considers how the two sides of finance (investing and lending) interact with economic, social, and environmental issues. Additionally, the European Commission (2020) describes sustainable finance as the process of taking due account of ESG issues in making investment decisions in the financial sector, which ultimately leads to increased longer-term investments in sustainable economic activities and projects.

Why has sustainable finance become a front burner issue in contemporary times? A number of studies provided cogent answers. The first reason why SF has become a thematic issue is because the commitment to achieving the SDGs requires a substantial investment annually to deliver the SDGs by 2030 (Franks et al., 2018; Sachs et al., 2019). Second, SF has become pervasive because its reaction to sustainability issues is escalating and growing among governments, companies, investors, customers, banks, and other financial institutions, as these critical stakeholders are compelled to integrate sustainability factors into their operations, management models, and governance frameworks (Calabrese et al., 2019; Rogers & Serafeim, 2019). Third, the effects of climate change and unsustainable practices on agriculture and food production supported the need to embrace SF for agricultural sector transformation. At present, scholars have noted that there is widespread discontent and escalating public pressure on agribusiness firms and other stakeholders in the agriculture value chain to embrace more sustainable management practices in agricultural production and food processing (Jansen & Vellema, 2004; Friedrich, Heyder, & Theuvsen, 2012). Fourth, sustainable finance is distinct from the traditional finance because it is a finance option that supports project investments, taking into account environmental, social, and governance considerations (Raimi et al., 2022).

Six popular sustainable finance options that have added value to the financial innovation experiment across the globe are green bonds, green credit, green investment funds, green insurance, impact bonds, and blue bonds (Raimi et al., 2022). These six sustainable financing options, as shown in Fig. 5.3, are all supportive of SDGs but have distinct differences. The green bond is a fixed-income instrument that is designed specifically to support climate-related or environmental projects. Green credit, by contrast, is a financing scheme offered by banks as innovative financing for supporting commercial projects and investment portfolios that strengthen sustainable development. A green investment fund is an investible fund that are offered to productive sectors and commercial activities that benefit the environment and comply with specific investment criteria, strategies, and processes. Green insurance is an eco-friendly type of insurance that provides insurance cover for projects and businesses against financial losses while contributing to protecting the ecosystem or environment. A fraction of the green insurance premium is donated to or invested in planting trees, conservation of rainforests, and climate smart agriculture. Impact bonds are a form of results-based financing whereby an investor provides upfront capital for public projects that deliver social and environmental outcomes. If the projects succeed, the investor is repaid with interest by the government, by a donor organization, or by a foundation. Blue bonds are an innovative financing instrument for supporting projects and infrastructural facilities on islands and in coastal nations with the aim of creating

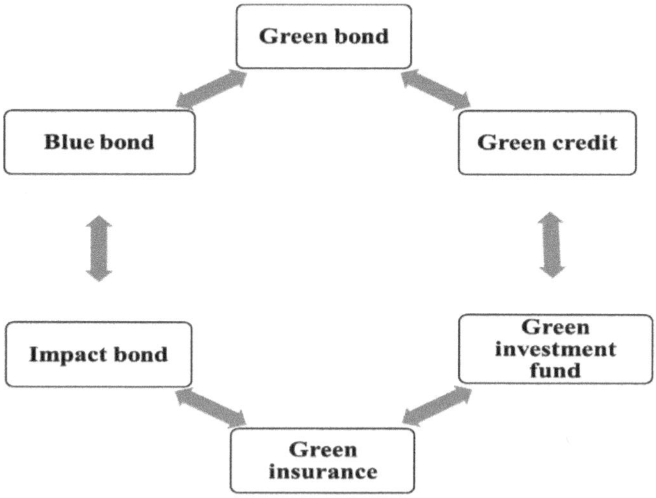

FIGURE 5.3 The six popular sustainable finance options. Source: *Developed by the authors.*

long-term sustainable financing for marine protection, sustainable economic development, and effects of climate change (Jeucken, 2010; Beschloss & Mashayekhi, 2019; Raimi, Shokunbi, & Olowo, 2022).

Beschloss and Mashayekhi (2019) noted that green financing instruments that have been developed to support sustainable projects and portfolios have been applauded by the Global Sustainable Investment Alliance because in 2018 alone, an estimated $30.7 trillion institutional assets were invested in sustainable green products in six markets, namely, the United States, Europe, Australia, Canada, Japan, and New Zealand. The prospect of green financial technologies is also being expounded because as an emerging digital-driven financial instrument, it has the propensity to accelerate the achievement of SDG 12 (responsible production) by promoting synergies between environmental SDG 1 and social SDG 15, which helps to increase agribusiness profitability without additional use of natural resources (Hinson, Lensink, & Mueller, 2019).

5.5 Capital markets union: an ambitious project for the development of emerging markets in central and eastern Europe

The European Capital Markets Union (CMU) is a bold project, with the Treaty of Rome being the first European document to address this goal. The creation of the common capital market was one of the objectives of the treaty that established the EU, and this issue was brought back to the attention of stakeholders in 2014 by the then-President of the European Commission, Jean-Claude Juncker. The attention of European officials turned to this project with the construction of the banking union (started in 2012), the aim being to increase the financing potential of the real economy that was offered by the capital market. Better capital mobilization (as a result of the implementation of specific measures) and targeting mainly small to medium enterprises (SMEs) and infrastructure projects are the key goals of this project (Friedrich & Thiemann, 2017; COM 292/2017). In contrast to the banking union, which targets only eurozone countries, CMU is a larger project that covers all EU countries (Coimbra Henriques, 2018). Therefore an integrated capital market of EU member states will support the process of growth, stability, and convergence through two key mechanisms: (1) facilitating the meeting between investors and companies or public authorities that are in need of financing (which promotes growth) and (2) improving private sector risk sharing (which supports economic stability).

Progress in the banking field regarding the harmonization of the legislative framework has ensured the smooth flow of capital movements at the European level, which is an asset in the process of removing other obstacles to the free movement of money. (Véron, 2014). The smooth functioning of the capital market union is based on a strong banking and fiscal union (Acharya & Steffen, 2016), given the importance of taxes in making cross-border investments in the financial market. Development of capital market-specific intermediaries and innovative instruments and techniques, such as public trading platforms for SMEs, private placements, venture capital funds, initial public offerings of shares, bond launch, securitization, and direct purchase of loans directly from banks by companies insurance and investment funds, is supported by European authorities in view of the progress that has been made in the U.S. capital market (Anderson et al., 2015):

- Improving access to finance for European companies, in particular for SMEs and investment projects such as infrastructure
- Diversifying and attracting funds from the economy and reducing the cost of capital
- Streamlining capital markets so that they are more efficient and effective

Despite the efforts of the European authorities, the CMU is still a desideratum on which the attention of the member states is focused. The establishment of a single market for capital movements has specific objectives that contribute to the development of the European Union, given the role of the financial market in the process of economic development. Despite significant progress in regional financial integration, EU capital markets are still fragmented, with significant gaps between them (Acharya & Steffen, 2016). This fragmentation significantly affects the size of the European capital market, and regional disparities between the north and south are substantial. These disparities make the European capital market far behind the U.S. and Asian markets (Gleisner & Thomadakis, 2018) in terms of specific indicators, such as market capitalization relative to GDP. Significant fluctuations are recorded annually for this indicator, given the significant impact that economic, social, and political factors have on stock market activity (Fig. 5.4).

Statistical data for eurozone countries show a similar evolution of the market capitalization/GDP ratio as for all EU economies, which shows the marginal position (Fig. 5.5) of Central and Eastern Europe (noneurozone countries). This can be explained by the late establishment (after 1990) of stock exchanges in these countries, which had abolished all institutions during the communist period, specifically the market economy.

According to consolidated data published by the Institute of European Capital Markets (ECMI, 2021) for 2019, at the EU level, there

FIGURE 5.4 Market capitalization/GDP ratio in the European Union (The World Federation of Exchanges). Source: *Statistics | The World Federation of Exchanges (world-exchanges.org) (world-exchanges.org).*

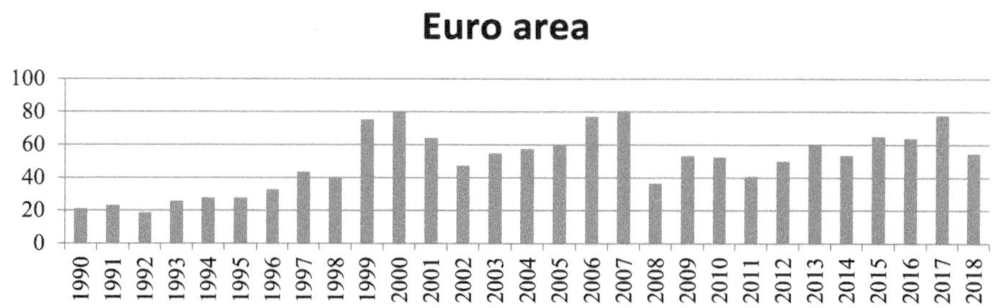

FIGURE 5.5 Market capitalization/GDP ratio in the eurozone. Source: *Statistics | The World Federation of Exchanges (world-exchanges.org) (world-exchanges.org).*

was 66%market capitalization relative to GDP, while this indicator had values of 321% in China, 162% in the United Kingdom, and 156% in the United States. Within the European Union the values of the indicator ranged from 155% in Sweden to 1% in Slovenia. The lowest values were recorded for the member countries of Central and Eastern Europe.

Statistical data published by the European Federation of Stock Exchanges show notable differences between European stock exchanges, with considerable gaps between West and East. Unfortunately, the exit of the United Kingdom from the European Union has led to a dramatic decline in financial activity at the regional level, given the position held by the London Stock Exchange (Fig. 5.6).

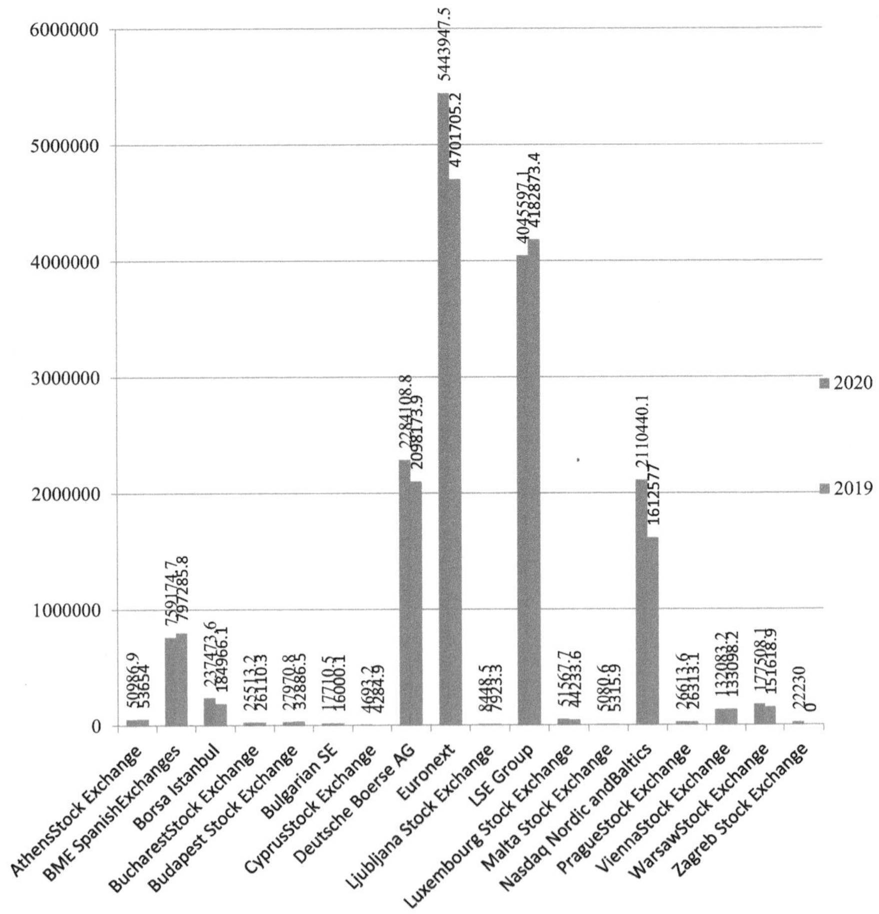

FIGURE 5.6 Market capitalization of the main European stock exchanges (in millions of U.S. dollars). Please see the online version to view the color image of the figure. Source: *Statistics | The World Federation of Exchanges (world-exchanges.org) (world-exchanges.org).*

One factor that has contributed significantly to the declining segmentation of European capital markets is the adoption of the euro (Bekaert et al., 2013). However, the financial crisis of 2009 led to a sharp decline in investor confidence in the capital market. Rebuilding investor confidence in capital market institutions and mechanisms is a cumbersome, time-consuming process because poor financial education mainly affects noneurozone countries (Ene, 2017; Castro et al., 2021).

Moreover, European capital markets are small in comparison to the US capital market, both in the equity component and in the bond segment. In addition, at the EU level, there is a worrying dependence on bank lending (80%), compared to only 20% in the United States (Commission 2015a). This reliance on indirect financing becomes even more dangerous, given the experience of the financial crisis of 2009. However, massive financing of European companies through market-specific mechanisms can, provide opportunities for capital market development by relaunching EU-wide securitization operations (Braun et al., 2018), which through additional regulations can regenerate the European securitization market, which must be reconfigured through transparency and stability. Therefore CMU offers a solution to increase financial security at the EU level, playing a complementary role to that of banks in the process of financing companies, especially SMEs.

Since the launch of the project, the completion of the single capital market has been subject to political pressures and divergences between financial market players: on one hand, universal banks, credit institutions in the London financial center, and the British authorities and, on the other, less enthusiastic hand, smaller banks, SMEs, and Central and Eastern European governments, where the capital market is in its early stages of development (Quaglia & Howarth, 2018). A certain reluctance is also manifested by countries whose financial systems are dominated by large banks that are owned by local capital (Germany, Italy, Austria), considering that CMUs promote the opening of financial markets and implicitly their denationalization. Brexit poses additional challenges for European authorities, which are thus losing the support and expertise of the British in the complex process of completing the single capital market.

Given the process of liberalization of capital movements at the international level and especially between the countries of the European Union, the establishment of a single capital market is an achievable goal. Moreover, the European authorities are aware of the need for such a union, given the long-term implications:

- Equalizing the cost of capital
- Raising capital and unlocking more investment in the EU but also in the rest of the world

- The emergence of financial investment opportunities for individual investors (including pension plans)
- Improving the connection between financing demand (EU-level investment projects) and the capital market-specific financing supply
- Increasing the stability of the European financial system
- Financial integration
- Reduction in financing costs
- Increasing the competitiveness of European companies

The Capital Markets Union "will support economic and monetary union by supporting economic convergence and contributing to the absorption of economic shocks in the euro area" (COM 468/2015). This capital market union will provide larger amounts of funds that can be accessed at lower costs, making SMEs the main beneficiaries given the difficulties and limitations that these entities face in accessing international financial markets. One solution for SMEs would be to relaunch securitization techniques, safely, given the potential of these techniques (Quaglia & Howarth, 2018). The European authorities' estimates are in excess of 100 billion euros, which could be attracted by banks and granted to companies through securitization operations (COM 468/2015). Reactivating venture capital funds would be another solution that would generate more than 90 billion euros to finance companies (Box 5.1).

This single capital market is expected to lead to an improvement in the resilience of the European financial system, which shows some shortcomings caused by the United Kingdom's exit from the European Union and COVID-19. The distribution of financial risks is ensured by cross-border economies for member countries. The benefits of the existence of the single capital market are felt by both institutional and individual investors.

The Capital Markets Union project aims to increase the financing opportunities of companies, particularly SMEs, as they are considered to be the backbone of the European Union (COM 590/2020). One solution is the Schemade Bank Recommendation, which was launched in the United Kingdom in 2016. Under this scheme, a bank that rejects an application for financing for SMEs over GBP 1000 must provide specific information that it holds in connection with the application to all financial platforms authorized by the Central Bank (after requesting the permission of the SME in question), thus supporting the financing of business. At the level of other European countries the solution that was adopted was simpler in the sense that the banks had to motivate the requested companies to identify the causes that led to the rejection of the financing application (Fig. 5.7).

In addition, through specific mechanisms the CMU can provide opportunities to implement specific EU economic policies, such as the

BOX 5.1

Securitization: an innovative financing technique (Altunbas et al., 2009; Shin, 2009)

Securitization of receivables is a form of financing through which financial assets that generate foreseeable future cash flows (e.g., mortgages, corporate loans, and auto loans) are grouped and transferred to a third party (a securitization fund), which finances this acquisition using funds attracted from the capital market through the issuance of securitized instruments and asset-backed securities. A typical securitization operation generally involves two steps:

- The transfer of assets (receivables) to the securitization fund
- The issuance by the securitization fund of asset-backed securities, which have the mortgage receivables behind them. The fund uses the money that is received from the issuance of securities to pay the original originator (bank) the value of the transferred assets.

The success of securitization operations in the US market was due to its advantages for credit institutions, investors, and other actors of capital market:

- It provides the necessary conditions for the development of new financing instruments for entities that provide the capital necessary to make acquisitions in the real estate market.
- It allows the development of a system that is meant to remove the blockages that could appear at some point in the future, owing to the lack of liquidity on the mortgage market.
- It offers creditors a new way to transform receivables into immediate liquidity and ensures an improvement of the risk management resulting from the granting of mortgage loans.
- It generates the emergence of new instruments with a significant potential to increase transactions on the securities market, which will ensure diversification of the portfolios of institutional investors.

However, specialists point out the devastating effects that securitization has had in triggering the financial crisis: "securitization is deemed to have been a major contributing factor to the 2007/08 financial crises by fueling credit growth accompanied by lower banks' credit standards" (Kara et al., 2016). Initially, it was believed that securitization would contribute to the stability and resilience of the financial system by transferring risk from the banking market to the capital market. In reality, the situation has degenerated as a result of multiple causes,

continued

BOX 5.1 (*cont'd*)

such as the complex nature of securitized financial products, the severe credit quality deterioration of financial instruments, and inadequate evaluation of these products by rating agencies (Senarath, 2017).

Public authorities, specialists, and capital market participants are aware of the importance of securitization as a source of financing, which is why efforts have been made in the European Union to promote this financing technique even after the financial crisis. It is currently strictly regulated by Regulation 2017/2402 on securitization

Source: Altunbas et al., 2009; Shin, 2009.

FIGURE 5.7 The main objectives of the European Union that can be achieved through the single capital market. Source: *Authors based on COM 590/2020 Communication from the commission to the European parliament, the council, the european economic and social committee and the committee of the regions a capital markets union for people and businesses-new action plan.*

Green Pact and the Digital Agenda. The European Union's transition to a low-carbon economy is a complex process involving not only technical and social challenges, but also investments of 200 billion euros a year (PRIMES, 2030 Impact Assessment). It is estimated that, compared to the previous decade, additional energy-related investments of 350 billion euros will be needed each year to meet the target of reducing greenhouse gas emissions by 55% by 2030. Some of these investments can be financed through capital market-specific mechanisms. One of the mechanisms with high potential for use is ESGs, also known as green bonds. The European authorities consider that green bonds are instruments through which the European Green Pact can be implemented. This is possible by standardizing these financial instruments on the basis of selection criteria of projects developed by international financial institutions, such as the World Bank, the EIB, and the European Bank for Reconstruction and Development, which have generated market liquidity and substantial growth of transactions by individual and institutional investors.

The CMU can also help to accelerate the recovery of EU countries from the COVID-19 crisis. Public authorities and companies will have additional funding opportunities as viable alternatives to bank funding. The need for funding from public authorities is growing in view of the economic and social recovery needs generated by the COVID-19 crisis and the necessary infrastructure projects at the European level. The COVID-19 crisis put pressure on the banking system, with bank credit being the main source of funding during this period. European authorities believe that the Capital Markets Union will offer financing alternatives and reduce its reliance on indirect financing specific to the banking sector.

Given the need to build a digital future, additional investment is needed to develop innovative businesses. The specificity of the activity of these companies and the lack of real guarantees generate their orientation toward financing mechanisms that are specific to the capital market. In addition, digitalization offers technical solutions that allow the development of the single capital market, with digital finance being one of the trends in Europe and worldwide (Fig. 5.8).

SMEs face major difficulties in accessing financial resources for the initiation and development of investment projects. The causes are multiple, including lack of a solid financial history, lack of material guarantees, and investors not having access to detailed information as is the case with large companies with a public character (i.e., listed on the stock exchange). Because bank credit covers 90% of the financing needs of SMEs, the aim is to diversify the sources of financing for these entities, especially as 18% of applicants have had financing applications rejected by credit institutions (COM 103/2020). Diversification of funding sources is necessary for the development of SMEs, and the proposed options include the following:

- Expanding venture capital funding through the ESCALAR initiative, which aims to increase the size of venture capital funds
- Creating a fund for the initial public offerings of SMEs to finance them on the capital market

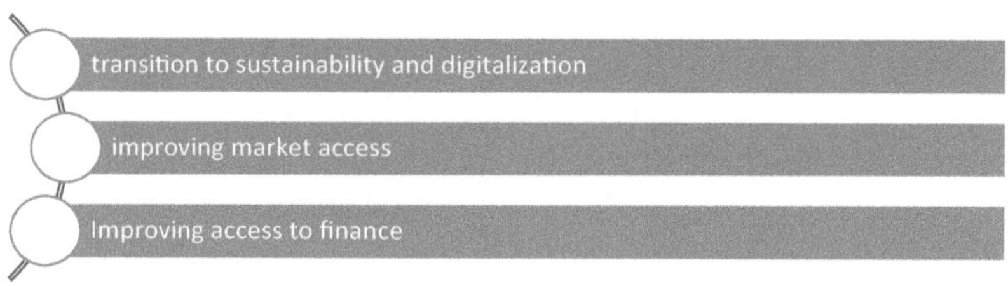

FIGURE 5.8 The pillars of the SME strategy for a sustainable and digital Europe. Source: *Authors based on SME strategy for a sustainable and digital Europe.*

- The use of blockchain technology that will allow the issuance of cryptographic assets and digital tokens in the form of bonds
- Smart funding from a gender perspective

The Digital Finance Strategy for Europe (COM 591/20200) is a bold document that aims to improve the digital performance of companies and consumers. Its main objectives are presented in Fig. 5.9.

Through the objectives that are set, this strategy will also have an impact on European capital markets by increasing the opportunities for investors to conduct cross-border transactions, fueled by the digitization of financial products and services. Retail investors have great potential to develop a single capital market because the EU savings rate is high. Extensive education campaigns launched by capital market institutions are needed to stimulate participation in the capital market, but also specific actions must be taken (Fig. 5.10).

Sustainable development and the recovery of the EU economy after the COVID-19 crisis are particularly complex processes that require the mobilization of significant long-term financial resources. Therefore specific strategies are needed that must be correlated to maximize results (Fig. 5.11).

The aging of the EU population is another challenge that authorities need to address (Vasile, 2012; Thalassinos et al., 2019; Cristea et al., 2020). Private pensions have the role of supplementing public pensions; therefore the single capital market must provide financial products and services that meet the needs of the population. Consumer financial

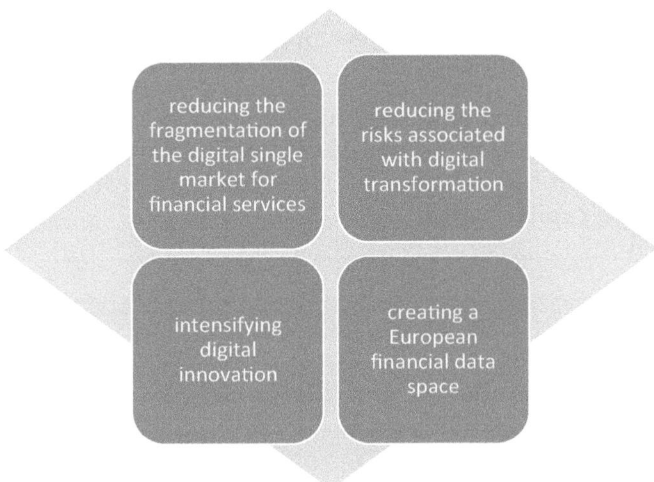

FIGURE 5.9 Objectives of the Digital Finance Strategy for Europe. Source: *Authors based on Digital Finance Strategy for Europe.*

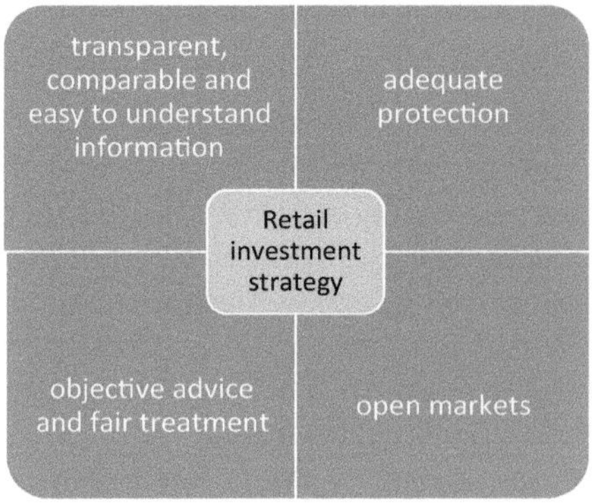

FIGURE 5.10 The main aims of retail investment strategy. Source: *Authors based on retail investment strategy.*

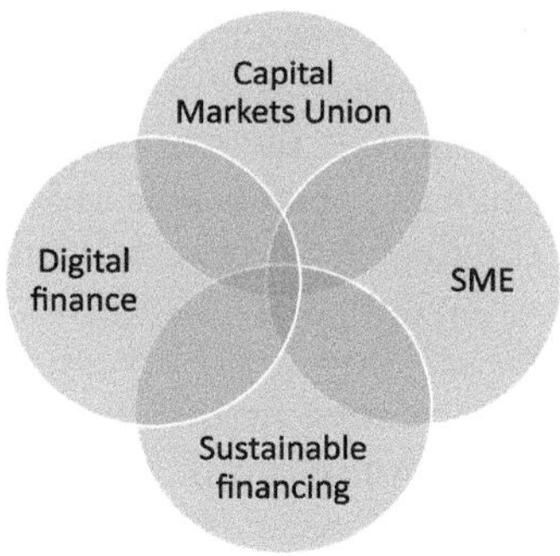

FIGURE 5.11 Strengthening the European Union through specific strategies aimed at interdependent strategies. Source: *Authors based on European Commission documents.*

education is essential, given the specific risks of the capital market. Moreover, pension plans have a long time horizon, which is why financial education programs need to be run for the active population who thus develop responsible behavior. In addition, specialized studies have highlighted the different impacts of financial education programs, depending on the age group to which they are addressed.

5.6 Conclusion, policy implications, and limitations

The capital market plays a favorable role in financing economic activity, with statistics showing the importance of this financial segment in the world's strongest economy: that of the United States. Based on this model, European research authorities should create a legal and institutional framework for the consolidation and convergence of European capital markets. This is necessary in view of certain considerations:

- The European capital market is fragmented as a result of the existence of national markets with particularities determined by the levels of development of each country, traditions, history, and so on.
- There is a need to offer SMEs alternative financing options to bank credit. For this reason, the European authorities not only encourage the financing of SMEs through capital market-specific instruments but also create the premises for a massive listing of them on the stock exchange by drawing up specific rules.
- The instruments of the secondary capital market are quintessentially speculative in nature, with the risk of becoming disruptive or even parasitic.
- The vector of business ethics is also a component of capital market investments, as many investors are interested in both the financial and social and environmental performance of listed companies.
- There is a need for public promotion of nonfinancial information, which is more urgent in light of the growing interest of investors in such information in order to substantiate capital market decisions.
- The need to increase the degree of information (financial information and information about the capital market) in view of the specific instruments that can be used by retail investors for making short-term and long-term investments, including pension plans (given the aging processof European population).
- Digitization will increase interest in transactions, especially from young people, but the positive effects on the development of the capital market are overshadowed by the risk of cyber-attack.

Despite the efforts that the European authorities have made to strengthen the single capital market, major differences can be seen between the old and the new states of the European Union. Despite regional cooperation and even regional mergers and acquisitions between exchanges (a fact favored by their demutualization), Central and Eastern European stock exchanges perform modestly in terms of the values of stock market transactions (and market capitalizations), the number of foreign companies, the supply of products financial, indexes released and listing conditions imposed (conditions on financial performance and more on social and environmental performance are imposed).

In view of the paucity of research on financial innovation for sustainable development in emerging markets, this chapter was written to fill in a major gap in the extant literature concerning catching up with sustainable development in emerging markets through financial innovation. In specific terms, the chapter explored sustainable financing tools, focusing on the impact of projects, programs, and portfolios and not only profit, emphasizing the intersection of finance and technology, financial innovation in the green economy, ecological issues, and women's inclusion. At the end of the discourse, a number of insightful facts emerged for policy and managerial considerations. First, the chapter contributes theoretically to and broadens the conceptual knowledge of financial market innovation through sustainable finance and other financial instruments. Therefore it would be in the interest of managers and professionals in financial markets to promote sustainable finance and other financial instruments for transition to a low-carbon economy. Second, the required innovative financial instruments in emerging markets intersect with the green economy, ecological issues, and women's inclusion for enduring and sustainable development. Third, in Eastern Europe, which is our research context, the concept of sustainable finance in financial markets entails the introduction of financial products and services, the development of new processes and structures for better interaction with customers and educating the public to encourage progression toward sustainable development, and social inclusion. Fourth, the popular sustainable finance options that have added value to the financial innovation experiment across the globe include green bonds, green credit, green investment funds, green insurance, impact bonds, and blue bonds. These six sustainable financing options are all supportive of SDGs but have distinct differences. To achieve sustainable development targets in emerging markets with less developed financial market operations and governance, this chapter advocated investing for positive social and environmental impacts alongside economic returns rather than investing purely for profit.

As useful as the insights from this chapter may be, there are a few limitations. The far-reaching insights that were explicated in the chapter are based on content analysis supported by descriptive statistics from emerging capital markets in Central and Eastern Europe. Both content and descriptive analyses cannot be used for making informed predictions and generalization. To overcome this limitation, future studies should consider using rigorous econometric tests such as the cointegration test, test of causality, and inferential statistics to analyze financial market datasets to enhance generalization and prediction. Our datasets are obviously limited in terms of financial data quality and diversity. We set out to looking at sustainable development in emerging markets through financial innovation, but the secondary data that were used were from emerging capital markets in Central and Eastern Europe. The insights from the chapter are therefore most applicable to Central and Eastern Europe and by extension to the European Union.

References

Acharya, V. V., & Steffen, S. (2016). Capital markets union in Europe: Why other unions must lead the way. *Swiss Journal of Economics and Statistics*, *152*(4), 319–329.

Ajide, F. M. (2016). Financial innovation and sustainable development in selected countries in West Africa. *Journal of Entrepreneurship, Management and Innovation*, *12*(3), 85–112.

Ajide, F. M., & Ajileye, J. O. (2015). Market concentration and profitability in Nigerian banking industry: Evidence from error correction modeling. *International Journal of Economics, Commerce and Management*, *3*(1), 1–8.

Alfaro, E., Yu, F., Rehman, N. U., Hysa, E., & Kabeya, P. K. (2019). *Strategic management of innovation. The Routledge companion to innovation management* (pp. 107–168). Routledge.

Allen, F. (2012). Trends in financial innovation and their welfare impact: An overview. *European Financial Management*, *18*(4), 493–514.

Altunbas, Y., Gambacorta, L., & Marques-Ibanez, D. (2009). Securitisation and the bank lending channel. *European Economic Review*, *53*(8), 996–1009.

Anderson, N., Brooke, M., Hume, M., & Kürtösiová, M. (2015). *Financial stability paper 33: A European capital markets union: Implications for growth and stability*. Bank of England. (No. 33).

Baregheh, A., Rowley, J., & Sambrook, S. (2009). Towards a multidisciplinary definition of innovation. *Management Decision*, *47*(8). Available from https://doi.org/10.1108/00251740910984578.

Beck, T., Tao, C., Chen, L., & Frank, M. S. (2016). Financial innovation: The bright and the dark sides. *Journal of Banking & Finance*, *72*, 28–51.

Bekaert, G., Harvey, C. R., Lundblad, C. T., & Siegel, S. (2013). The European Union, the Euro, and equity market integration. *Journal of Financial Economics*, *109*(3), 583–603.

Beschloss, A., & Mashayekhi, M. (2019). Point of view–a greener future for finance: Green bonds offer lessons for sustainable finance. *Finance & Development*, *56*(004).

Błach, J. (2020). Barriers to financial innovation—corporate finance perspective. *Journal of Risk and Financial Management*, *13*(11), 273.

Braun, B., Gabor, D., & Hübner, M. (2018). Governing through financial markets: Towards a critical political economy of capital markets union. *Competition & Change*, *22*(2), 101–116.

Brezeanu, P., Şimon, I., & Novac, L. E. (2005). *Instituţii financiare internaţionale*. Editura Economică.

Brezoi, A. G. (2018). Ethics and corporate social responsibility in the current geopolitical context. *Economic Insights–Trends and Challenges*, *7*, 45–52.

Calabrese, A., Costa, R., Levialdi, N., & Menichini, T. (2019). Integrating sustainability into strategic decision-making: A fuzzy AHP method for the selection of relevant sustainability issues. *Technological Forecasting and Social Change*, *139*, 155–168.

Çankaya. (2014). Pros and cons of financial innovation. In H. Dincer, & Ü. Hacioglu (Eds.), *Globalization of financial institutions*. Cham: Springer. Available from https://doi.org/10.1007/978-3-319-01125-7_12.

Castro, R. G., Enríquez-Diaz, J., & García, B. A. (2021). The role of financial education in the path towards sustainable development: A practical experience with students of secondary education. In *Financial management and risk analysis strategies for business sustainability* (pp. 1–19). IGI Global.

Coimbra Henriques, S. (2018). The role of the capital markets union: Towards regulatory harmonisation and supervisory convergence. *Perspectives on Federalism*, *10*(1).

COM 103/2020 *Communication from the commission to the european parliament, the council, the European economic and social committee and the committee of the regions an SME strategy for a sustainable and digital Europe.*

COM 468/2015 *Communication from the commission to the european parliament, the council, the european economic and social committee and the committee of the regions action plan on building a capital markets union.*

COM 590/2020 *Communication from the commission to the European parliament, the council, the european economic and social committee and the committee of the regions a capital markets union for people and businesses-new action plan.*

COM(2017) 292 *Final comunicare a comisiei către parlamentul European, consiliu, comitetul economic Şi social European Şi comitetul regiunilor referitoare la evaluarea la jumătatea perioadei a Planului de acţiune privind uniunea pieţelor de capital.*

COM(2020) 591 *Communication from the commission to the european parliament, the council, the European economic and social committee and the committee of the regions – a digital finance strategy for the EU.*

Commission (2015a) *Green paper: Building a capital markets union.* COM(2015) 63 final, 18 February.

Cristea, M., Georgiana Noja, G., Dănăcică, D. E., & Ştefea, P. (2020). Population ageing, labour productivity and economic welfare in the European Union. *Economic Research-Ekonomska istraživanja, 33*(1), 1354–1376.

Denscombe, M. (2017). *Ebook: The good research guide: for small-scale social research projects.* (UK): McGraw-Hill Education.

Dirk S., (2018): *A framework for sustainable finance, article in SSRN electronic journal January 2018.* <https://www.researchgate.net/publication/323764756>.

Duffie, D. (2014). Financial market infrastructure. *Across the Great Divide: New Perspectives on the Financial Crisis, 652,* 251.

ECMI. (2021). *European Capital Markets Institute* | European Capital Markets Institute (ECMI)

Ene, C. (2017). Current issues regarding the protection of retail investors on the capital market within the european union. *The USV Annals of Economics and Public Administration, 17*(1 (25)), 35–44.

European Commission. (2020) *What is sustainable finance?* European Commission website. <https://ec.europa.eu/info/business-economy-euro/banking-and-finance/sustainable-finance/overview-sustainable-finance_en>.

Evstigneev, I., Hens, T., & Reiner Shenk-Hope, K. (2004). *Evolutionary finance (National Center competence research financial valuation and risk management.* Working Paper No. 462).

Frame, W. S., & White, L. J. (2014). *Technological change, financial innovation, and diffusion in banking* (pp. 1–5). Leonard N. Stern School of Business, Department of Economics.

Franks, M., Lessmann, K., Jakob, M., Steckel, J. C., & Edenhofer, O. (2018). Mobilizing domestic resources for the Agenda 2030 via carbon pricing. *Nature Sustainability, 1*(7), 350–357.

Friedrich, J., & Thiemann, M. (2017). Capital markets union: The need for common laws and common supervision. *Vierteljahrshefte zur Wirtschaftsforschung, 86*(2), 61–75.

Friedrich, N., Heyder, M., & Theuvsen, L. (2012). *Sustainability management in agribusiness: Challenges, concepts, responsibilities and performance* (No. 1020–2016–81734, pp. 530–546).

Gigauri, I. (2021). Corporate social responsibility and COVID-19 pandemic crisis. *International Journal of Sustainable Entrepreneurship and Corporate Social Responsibility, 6* (1), 30–47. Available from https://doi.org/10.4018/ijsecsr.2021010103.

Gigauri, I., & Djakeli, K. (2021). Expecting transformation of marketing during the post-pandemic new normal. *International Journal of Sustainable Economies Management, 10*(2), 1–18. Available from https://doi.org/10.4018/ijsem.2021040101.

Gleisner, L., & Thomadakis, A. (2018) *Recent developments in European capital markets: Key findings from the 2018 ECMI Statistical Package,* Research Report No. 14 December 2018

Hinson, R., Lensink, R., & Mueller, A. (2019). Transforming agribusiness in developing countries: SDGs and the role of FinTech. *Current Opinion in Environmental Sustainability, 41,* 1–9.

Hysa, E., Kruja, A., Rehman, N. U., & Laurenti, R. (2020). Circular economy innovation and environmental sustainability impact on economic growth: An integrated model for sustainable development. *Sustainability, 12*(12), 4831.

Iacovoiu, V., & Stancu, A. (2017). An empirical analyse of the correlation between innovation performance and economic development. *The USV Annals of Economics and Public Administration, 17*(1 (25)), 45–51.

Ionescu, R. (2019). Corporate social responsibility programs of the bucharest stock exchange. *Economic Insights-Trends and Challenges, VIII(LXXI)*(3), 45–54.

Jansen, K., & Vellema, S. (Eds.), (2004). *Agribusiness and society: Corporate responses to environmentalism, market opportunities and public regulation*. London: Zed Books.

Jeucken, M. (2010). *Sustainable finance and banking: The financial sector and the future of the planet*. Routledge.

Johnson, N. F., Jefferies, P., & Hui, P. M. (2003). *Financial market complexity. OUP catalogue*. Oxford University Press, number 9780198526650.

Kapoor, R., & Mention, A. L. (2015). Patenting financial innovation in Europe. In *2015 Portland International Conference on Management of Engineering and Technology (PICMET)* (pp. 1071–1077). IEEE.

Kara, A., Ozkan, A., & Altunbas, Y. (2016). Securitisation and banking risk: What do we know so far? *Review of Behavioral Finance, 8*(1), 2–16.

Khan, S. A. R., Yu, Z., Panait, M., Janjua, L. R., & Shah, A. (Eds.), (2021). *Global corporate social responsibility initiatives for reluctant businesses*. IGI Global.

Laeven, L., Levine, R., & Michalopoulos, S. (2015). Financial innovation and endogenous growth. *Journal of Financial Intermediation, 24*(1), 1–24.

Lerner, J., & Tufano, P. (2011). The consequences of financial innovation: A counterfactual research agenda. *Annual Review of Financial Economics, 3*(1), 41–85.

Llewellyn, D. T. (2009). Financial innovation and the economics of banking and the financial system. In L. Anderloni, & D. T. Llewellyn (Eds.), *Financial* innovation in retail and corporate *banking*. Cheltenham: Edward Elgar.

Marszk, A., & Lechman, E. (Eds.), (2021). *The digitalization of financial markets: The socioeconomic impact of financial technologies*. Routledge.

Matei, M. (2013). *Responsabilitatea socială a corporaţiilor şi instituţiilor şi dezvoltarea durabilă a României*. Bucharest: Expert Publishing House.

Mention, A. L. (2011). Innovation for financial services. *Innovation Management*. <http://www.innovationmanagement.se/2011/09/13/innovation-for-financial-services>.

Merton, R. C. (1995). Financial innovation and the management and regulation of financial institutions. *Journal of Banking & Finance, 19*(3–4), 461–481.

Michalopoulos, S., Laeven, L., & Levine, R. (2009). *Financial innovation and endogenous growth* (No. w15356). National Bureau of Economic Research.

Motta, M. (2004). *Competition policy: Theory and practice*. New York: Cambridge University Press.

Nesvetailova, A. (2014). Innovations, fragility and complexity: Understanding the power of finance. *Government and Opposition, 49*(3), 542–568.

Nickell, S. J. (1996). Competition and corporate performance. *Journal of Political Economy, 104*(4), 724–746.

Panait, M., Voica, M. C., & Radulescu, I. (2014). The activity of capital market'actors: Under the sign of social responsibility. *Procedia Economics and Finance, 8*, 522–528.

PRIMES. (2030) *Impact assessment*.

Pyka, I. (2013). Produktowe innowacje banków komercyjnych: Analiza przyczynowo-skutkowa. *Studia Ekonomiczne, 173*, 253–267.

Quaglia, L., & Howarth, D. (2018). The policy narratives of European capital markets union. *Journal of European Public Policy, 25*(7), 990–1009.

Raimi, L., Shokunbi, M. O., & Olowo, R. (2022). Rethinking the prospects of sustainable finance and challenges of agribusiness transformation in Nigeria: Implications of the nexus for entrepreneurship development. In I. Yasar Gok (Ed.), *Handbook of research on global aspects of sustainable finance in times of crises* (pp. 312–332). IGI Global. Available from http://doi:10.4018/978-1-7998-8501-6.ch015.

Răzvan, I. (2021). Financial literacy in the digital age: Challenges and opportunities in the European union. *Revista Estrategia Organizacional, 10*(2).

Rogers, J., & Serafeim, G. (2019). *Pathways to materiality: How sustainability issues become financially material to corporations and their investors* (pp. 20–056). Harvard Business School.

Sachs, J. D., Woo, W. T., Yoshino, N., & Taghizadeh-Hesary, F. (2019). Importance of green finance for achieving sustainable development goals and energy security. *Handbook of Green Finance: Energy Security and Sustainable Development*, 1–10.

Salas, O. (2009). *The future of innovation: A leverage for sustainable development and social inclusion.* <http://thefutureofinnovation.org/contributions/view/598/the_future_of_innovation_a_leverage_for_sustainablt_development_and_social_inclusion>.

Saunders, M., Lewis, P., & Thornhill, A. (2016). *Research methods for business students* (7th ed.). Harlow: Pearson Education.

Senarath, S. (2017). Securitisation and the global financial crisis: Can risk retention prevent another crisis? *International Journal of Business and Globalisation, 18*(2), 153–166.

Shiller, R. J. (2013). Capitalism and financial innovation. *Financial Analysts Journal, 69*(1), 21–25.

Shin, H. S. (2009). Securitisation and financial stability. *The Economic Journal, 119*(536), 309–332.

Sommer, S. (2020). *Sustainable finance: An overview.* Deutsche Gesellschaft für Internationale Zusammenarbeit (GIZ). GmbH, Bonn and Eschborn.

Stancu, A., & Iacovoiu, V. (2015). An econometrical approach of the relationship between innovation and net outward investment position. *The USV Annals of Economics and Public Administration, 15*(2 (22)), 7–18.

Stancu, S., Bodea, C. N., Naghi, L. E., Popescu, O. M., & Neamtu, A. (2017). *Use of new innovative technologies in business by all age groups. Digital entrepreneurship and global innovation* (pp. 79–103). IGI Global.

Thalassinos, E., Cristea, M., & Noja, G. G. (2019). Measuring active ageing within the European Union: implications on economic development. *Equilibrium, 14*(4), 591–609.

The Economic Times. (2021, December) *Definition of 'Capital Market.* Bennett, Coleman & Co. Ltd. <https://economictimes.indiatimes.com/definition/capital-market>.

The World Federation of Exchanges (world-exchanges.org)

Vasile, V. (2012). A new model for old age insurance. The private pension and financial sustenability. *Business & Leadership, 10*(2).

Vasile, V., Panait, M., & Apostu, S. A. (2021). Financial inclusion paradigm shift in the postpandemic period. Digital-divide and gender gap. *International Journal of Environmental Research and Public Health, 18*(20), 10938.

Véron, N. (2014). *Defining Europe's capital markets union* (No. 2014/12). Bruegel Policy Contribution.

Voica, M. C., & Stancu, A. (2021). Corporate social responsibility reporting: Background, evolution and sustainability promoter. *Sustainable Management for Managers and Engineers*, 109–155.

Williams, T. A., & Shepherd, D. A. (2017). Mixed method social network analysis: Combining inductive concept development, content analysis, and secondary data for quantitative analysis. *Organizational Research Methods, 20*(2), 268–298.

CHAPTER

6

Corporate social responsibility, business ethics, sustainability, and innovation: a theoretical analysis of the importance of these concepts in times of pandemic

Carla Marisa Rebelo de Magalhães

Transdisciplinary Research Center of Innovation & Entrepreneurship Ecosystems, Lusófona University, Porto, Portugal

6.1 Introduction

In recent decades there has been increasing concern worldwide with issues such as sustainable development because of the impact that organizational activity has on the environment and, consequently, because of the pressure that society has been exerting for companies to assume their social responsibilities and act ethically. Therefore a lot has been done in this field, including the use of cleaner technologies, awareness of issues such as the ecological and carbon footprint that each organization promotes, and regulations penalizing those who do not comply with certain environmental requirements. However, there is still a long way to go in this matter because we still do not have the necessary degree of commitment from many organizations and many countries, which suggests that regulation and awareness raising must be more rigorous and effective.

In fact, OECD suggests that if nothing is done in government terms, the global emission of greenhouse gases could increase by around 70% by

Innovation and Sustainable Manufacturing
DOI: https://doi.org/10.1016/B978-0-12-819513-0.00006-8

2050. It is therefore urgent that countries focus on this matter and prioritize laws that not only promote the sustainability and innovation of the business fabric, but also send a clear message that there can no longer be tolerance for polluting technologies and the lack of environmental awareness. Besides that, sustainability is not just an environmental, social, or political issue, but also an economic one, as it can promote the organizations' productivity, through the more efficient use of resources.

This theme becomes even more relevant in the current context of the pandemic, marked by constant uncertainty, which makes it urgent and almost mandatory for organizations to worry about issues such as innovation and sustainable management as a matter of survival. Indeed, the responses to COVID-19 must undergo a paradigm shift based on planned short-term strategies; real, effective, and targeted incentives for innovation; solid commitments to social responsibilities by organizations; and investments in more efficient technologies that promote not only sustainable management, but also organizational competitiveness. Otherwise, we may ask the following questions: What is the importance of innovation if it is not based on practices that support the organization's activity with a long-term perspective? Can we really talk about innovation if sustainable management is not considered? How can companies combine these concepts and make them the engine of progress and value creation?

We will reflect on these issues in this chapter. For that, it will be important to have as a starting point four key concepts that, acting in an integrated manner, can promote sustainable management: corporate social responsibility (CSR), business ethics, sustainability, and innovation.

6.2 Corporate social responsibility, business ethics, sustainability, and innovation: conceptualization and correlation

One of the most important concepts for sustainable innovation is the concept of CSR, which has become increasingly important in the literature, especially since the 1960s (Paul & Parra, 2021). However, this is not a consensual concept. The conceptual diversity of CSR can be observed through the work of Garriga and Melé (2004), who presented four major theories in this area: instrumental, political, integrative, and ethical.

Let's start by identifying some of the main authors who of instrumental theories. The authors who stand out the most in this regard are Levitt and Friedman. According to Levitt (1958), financial development is the basis of the capitalist system, and that is why companies should focus on it; anything that is CSR that goes beyond that should not be considered. Friedman (1962) agrees with this position, stating that companies should engage with CSR only if it is profitable. This author makes use of Adam Smith's idea, in the sense that the company should only seek profit, since an "invisible hand" will take care of the rest. Friedman (1970) also states that CSR, as defended by many authors, is an anti-profit-maximizing behavior that is assumed to benefit others than the company's shareholders, that is, such practice is not in accordance with an economy market, representing a burden for the company and an additional cost that reduces its effectiveness regarding targeted profits.

McWilliams and Siegel (2001), in turn, warn that there is an optimal point of application of resources in social actions, after which the company loses money. The authors look at CSR as a differentiation, as consumers can value such actions if they are aware of them. Therefore the positive perception of these practices adds value to a brand, allowing the company to charge more for its products and services. The question is to identify how far it makes sense to increase costs to achieve these improvements and to communicate them to the market. In this context, according to the authors, decisions involving CSR should receive the same treatment as any other investment decision, which is rarely done. For the authors there is a lack of consistency in empirical CSR studies, which is associated with the lack of a theory that links the issue to market forces, which is also identified by Marrewijk (2003) when he states that CSR has a lack of clarity in its theoretical principles.

In the perspective of Barry (2002), the belief that the company has other obligations in addition to those that focus on its owners, forces it to deal with a vague entity, namely, society, which ends up being represented by pressure groups that do not always have convergent or even viable needs.

Turning now to political theories, these advocate that it is through the CSR that companies relate to society and assume their power and political positions. In this perspective we can place Bowen, for whom CSR is defined as a set of obligations of businesspeople to pursue policies that are important in terms of the goals and values of our society (Bowen, 1953). The work of this author is based on the belief that large companies are vital centers of power and decision making and that the actions of these companies are at the base of their relationship with citizens.

Several authors refer to the concept of corporate citizenship as a concept that goes beyond CSR and is related to the impact that companies have on society and how they manage this impact and relate to the environment in which they act, demonstrating influence and power (Hemphill, 2004; Maignan & Ferrell, 2000, 2001a, 2001b; Matten, Crane, & Chapple, 2003; Peterson, 2004; Waddock & Smith, 2000; Warhurst, 2001; Wood & Logsdon, 2002).

Turning now to integrative theories, these state that companies must consider the demands and needs of society in their decision-making processes, as their survival depends on social will. Carroll argues that the CSR of a business involves the economic, legal, ethical, and discretionary expectations that society has in each period (Carroll, 1979). This definition has four components: The company is expected to produce goods and services and sell them at a profit (the economic component), comply with the law (the legal component), have ethical behavior and comply with standards (the ethical component), and voluntarily play roles that benefit society (the discretionary or philanthropic component). These four components are not mutually exclusive but have an accumulation character, from an economic base to a discretionary or philanthropic status. However, after working with Mark Schwartz, Carroll understood that the vertical formatting of this model does not capture the interrelationship between different levels and can convey a notion of priority of some levels over others, which is incorrect. Furthermore, the author asserts that the discretionary level is unnecessary and cannot be a business responsibility. Moreover, this dimension is already included in the ethical guidelines. Hence Carroll proposes another model, which, instead of being pyramidal and having four components, is circular and includes only three components (economic, legal, and ethical), which intersect with each other, with none prevailing over the others, but an overlap (Carroll & Schwartz, 2003).

Another author who is concerned with the issue of CSR from an integrative perspective (albeit in a gradual perspective) is Frederick (1986, 1994, 1998). The author's theory presupposes a sequence that systematically deepens the company's concerns. At the first level, named CSR1 by the author, is the idea that CSR promotes an obligation to improve social issues (Frederick, 1986, 1994). This obligation is based on work in the economic, social, and political fields and responds to the needs of the social fabric. Its origin can be found in the 1950s, and it prevailed until the mid-1970s, always linked to studies called business-and-society.

The evolution of this concept, according to Frederick (1994), takes it to the level of CSR2, or corporate social responsiveness, which is the ability of a company to face and answer social pressures. Therefore CSR2 means the manager's ability to conduct a company's relationships with the several social groups that involve it and is based on management practices as they concern practical aspects of the company's relationship with its environment.

The other level proposed by Frederick (1986) is CSR3, a concept that promotes the notion of good moral behavior in the context of the organization. At this level, the company must accept ethics as belonging to its core and decision making. For this, the company must train its managers to accept and practice the concept in their daily lives and provide analytical tools that are capable of anticipating ethical problems that affect the company and its employees. The dissemination of a culture that values ethical aspects is essential for the success of the company.

Finally, the last level proposed by Frederick (1998) is CSR4, which integrates three elements: the cosmos (C), science (S), and religion (R).

Also in this field, we can refer to the stakeholder theory, which is perhaps the most mentioned in the CSR area and states that all parties that have an interest in the company's activity should be considered (e.g., employees, customers, suppliers, government, nongovernmental organizations, competitors, distributors, and the community) and not just its shareholders (Evan & Freeman, 1993). This idea has been promoted by other authors (Freeman, 1984; Phillips et al., 2003; Stoney & Winstanley, 2001; Vos, 2003).

In the works of Rhenman and Stymne (1965) in Sweden and of Ansoff, in the United States, both in the 1960s, Carroll (1998) found the roots for the concept of stakeholder. According to Rhenman and Stymne, there were conflicts of interest between the different stakeholders, with insufficient means to satisfy all these interests, which depended on the capacities and orientations of the managers.

However, it was with Freeman that the concept of stakeholder became popular. According to the author, the stakeholder theory would be founded on the company's capacity to be influenced and, in turn, also to influence several actors in its environment, above all suppliers, shareholders, distributors, and executives (Freeman, 1984). Thus a stakeholder would be any group or individual that affects or is affected by the achievement of the company's objectives, which is the definition that has become most popular.

Stoney and Winstanley (2001) determined that the concept of stakeholder emerged in the 1960s, from the proposals of researchers at the Stanford Research Institute (SRI). These academics defined an alternative to the hegemonic path, which looked only at the interests of shareholders; this alternative path considered it important to extend the company's responsibility to a variety of stakeholders, without whom no organization could exist.

Freeman and Reed (1983) also stated that the term *stakeholder* was formalized by the SRI in 1963 based on an internal memorandum that referred to the groups without which organizations would cease to exist. This question was addressed, throughout the 1970s, by the Harvard Business School, which embraced the CSR project. Later, in 1977 the Wharton School initiated, at its Applied Research Center, a stakeholder project (Freeman & Reed, 1983).

Thus we see that there is a great variety in the orientations that mark the beginning of the concern with the stakeholder theory. Nevertheless, the objective is common: to extend the organizations' responsibilities toward other actors that go beyond the shareholders. But in this view, who are these stakeholders?

For Wilson there is a certain consensus regarding some stakeholders. Thus it is possible to identify, in the work of most authors on the subject, the presence of employees, consumers, shareholders, and suppliers as those to which a company should pay more attention (Wilson, 2003). Beyond these groups, there is greater difficulty in reaching full agreement. Also, according to Wilson (2003), there are authors who understand that stakeholders are all those who run risks as a result of the company's activities. Another group of thinkers understands that everyone who suffers the impacts of the action of a certain company can be classified as a stakeholder, whether the impacts are of a climatic, marketing, cultural, or other nature.

For Freeman and Reed (1983) the issue can be seen in a broader or more restricted sense. In a broader sense, stakeholders comprise all identifiable groups or individuals that can affect an organization in achieving its corporate goals. In this case, there are government groups, trade associations, competitors, unions, employees, customer segments, shareholders, and others. From a more restricted point of view, only those groups that the company needs to survive are considered. They are employees, customer segments, some suppliers, key government agencies, shareholders, some financial institutions, and others. Freeman (1984) also points to a group of primary stakeholders, which maintain a formal, contractual, and official relationship with the company. All others are classified as secondary stakeholders.

Phillips et al. (2003) present a more restricted definition, which they call normative, that encompasses funders, employees, customers, suppliers, and the local community, and a broader one, which also includes competitors, terrorists, activists, and the press.

Carroll (1991) understands as stakeholder the groups that have an interest in the company's operations and decisions and therefore have interdependencies with it. In a broad sense, such groups or individuals are shareholders, consumers, employees, suppliers, the community, and social activist groups.

Thus there seems to be a growing consensus regarding the fact that organizations should not consider only the interests of their shareholders and that the most correct—that is, the more ethical—stance is to consider all of the company's stakeholders.

Ethical theories, which advocate that the actions of companies should be based on ethical and moral values, rather than economic ones, exemplify the approach that most opposes the instrumental view.

Enderle and Tavis (1998) presented a model in which three levels of ethical challenges relate to corporate responsibilities in the economic, social, and environmental dimensions: level 1 (minimum ethical requirements), level 2 (obligations beyond the minimum ethical level), and level 3 (aspirations for ethical ideals). The conduct of each corporation can be analyzed in terms of the stage at which the corporation finds itself, according to these three levels, in each of the dimensions of corporate responsibilities: social, economic, and environmental. None of these dimensions of responsibility can be instrumentalized over the others, and each of them must meet minimum ethical requirements.

Kok, Van der Weile, McKenna, and Brown (2001) state that companies must be socially responsible to society because it is the correct thing to do and not because the companies want to receive something in return.

Mintzberg, Simons, and Kunal (2002) state that companies should focus on ethical values and not just on economic, environmental, and social issues. This is a vision that calls for social justice.

Reidenbach and Robin (1988) identified five major philosophical lines and isolated them in a model. This model of analysis of ethical behavior in business was later validated by the results of subsequent research carried out by the authors themselves (Reidenbach & Robin, 1990) and later by Hansen (1992), Cohen, Pant, and Sharp (1993).

The philosophical currents that were identified by Reidenbach and Robin (1988) are relativism, egoism, utilitarianism (teleology), deontology, and justice. These five dimensions of the Reidenbach and Robin model do not appear isolated, independent of each other. The research of Hansen (1992) and Cohen et al. (1993) show that they are present, all at the same time, in the ethical content of a behavior or a decision. They are ethical structures that compete on the basis of high-level moral and behavioral values and that determine ethical behavior in decision making. Let's look at each of them in more detail.

Starting with relativism, according to Reidenbach and Robin (1990), the basic concept of relativism is that all normative beliefs are a function of culture or the individual, so no universal rule for ethics can be applied to everyone. Thus no single and universal criterion can be used to determine when an action is or is not ethical. This concept even extends to a broader generalization about culture, in which ethical rules that define a person's behavior and values in one culture do not necessarily define them in another. For the study of business ethics the individual aspect of this theory is used in greater depth because there are personal differences between individuals or between individuals and society.

In relation to egoism, this ethical current is used in theories that measure the morality of things in the consequences of actions. Thus egoism shows that the individual guides his or her decisions so that the consequences are in his or her favor. The only valid standard for a person's behavior is for a person to stand above others (Reidenbach & Robin, 1990).

Regarding utilitarianism as well as selfishness, ethical behavior is also evaluated by the consequences it causes. However, for utilitarianism, what matters is that the result of this behavior benefits the greatest number of people possible in a society (Reidenbach & Robin, 1990).

The ethical current of deontology refers to the specific actions or obligations of individuals. It suggests that people base their attitudes on obligations to society and to other people, satisfying their legal requirements. Deontology is concerned with the justice that is in action or behavior (Reidenbach & Robin, 1990).

As for justice, Reidenbach and Robin (1990) suggest that this current encompasses many individual philosophical lines, but they depart from the idea of justice that was defined by Aristotle in his work "The Principle of Formal Justice." Justice preaches an equal distribution of benefits for everyone and implies that everyone should be treated equally. Although justice thus seems universal, it is not simple to define what criteria will be chosen so that it is possible to know what is equal for all or how a benefit can be distributed equally to all. This is known as distributive justice and is widely used today in business ethics discussion.

Regarding business ethics, several authors have dedicated themselves to its study, stating that it has been accentuated since 1970 (Salehi, Saeidinia, & Aghaei, 2012). Other authors state that its origins date back to the 19th century, when the first references on the subject appeared. From that point onward, the theme of ethics began to proliferate in the management practices of companies.

Despite its prominent position in the business world, especially through CSR, business ethics is not always welcomed. According to some authors, there seems to be a clash of interests between the activities that carry out the search of profit and compliance with ethical assumptions within organizations (Sundman, 2000).

Reference should also be made to the existing distances between the discourses, their codes, and the ethical behavior of organizations and their administrators. At this level, it is important to introduce the question of codes of ethics. Indeed, to be ethical, it is not enough to create a code of ethics. If the culture is not driven by values or reasons of a substantive nature, codes of ethics serve only to improve the company's image, even though they often can give employees a certain incentive to act ethically or can act as a brake when the company is faced with certain unethical behaviors because of the psychological effect that their exposure transmits. Perhaps that is why there are authors who consider the creation of a code of ethics to be always positive, even though the ideal is that it is incorporated into the company's management policy, including all employees from the bottom to the top of the organization (Schawartz, 2002). To this end, ethical leadership is essential so that ethical behavior is disseminated among employees (Halbusi, Ismail, & Omar, 2021) and is promoted in a sustainable manner.

Indeed, the concept of ethics ends up transporting us to the theme of sustainability. Over the past two decades, the concept of sustainability has attracted academics and professionals around the world. Incorporating three pillars (Ranjbari, Morales-Alonso, Shams Esfandabadi, & Carrasco-Gallego, 2019), sustainability focuses on social, environmental, and economic performance, with a view to the future of generations to come (Geissdoerfer, Savaget, Bocken, & Hultink, 2017). Because of the ambiguity that has emerged from sustainability, the triple bottom line (TBL) concept was proposed by Elkington (1998) in an attempt to objectify and contextualize the mentioned three pillars of sustainable development (Gimenez, Sierra, & Rodon, 2012).

In fact, companies and their stakeholders increasingly show a greater focus on and interest in exploring the development of sustainability (Hahn, Figge, Aragón-Correa, & Sharma, 2015) based on the notion that companies, in addition to focusing on profit, have a responsibility to generate environmental and social benefits (McWilliams, Parhankangas, Coupet, Welch, & Barnum, 2016). The focus of sustainable management is the performance of companies related to the TBL concept (Gawel, 2012; Gupta & Kumar, 2013). The adoption of sustainability initiatives can provide companies with several forms of benefits, including a positive image, greater trust from their stakeholders, greater efficiency in resource management, a competitive advantage, and superior returns on investments (Chen, 2010; Dangelico & Pujari, 2010).

Why do organizations engage with sustainability initiatives? This is a question that has been explored by some authors (Horak, Arya, & Ismail, 2018). We can point to the legitimacy of business performance, business success, and improvement in internal processes (Windolph, Harms, & Schaltegger, 2014). Indeed, acquiring or maintaining legitimacy has been described as the main motivating factor for sustainability (Scott, 2014). Black and Härtel (2004) highlight, as a driver of this legitimacy, the community values

that are influenced by the regulations with which organizations need to comply to remain legitimate. Johansen and Nielsen (2012) consider sustainability to be a type of narrative that organizations engage in, driven by isomorphic pressures toward achieving legitimacy as the main concern.

Regarding business success, the main motivation is related to consumers, who reward organizations by buying their products. Investors also can benefit organizations through their investment (Babiak & Trendafilova, 2011; Ditlev-Simonsen & Midttun, 2011). In fact, considering sustainability issues in product development can result in increased demand for the products (Dunphy, Griffiths, & Benn, 2007).

In the case of improving internal processes, it is known that the optimization of processes throughout the production chain can lead to a reduction in costs and consumption of certain resources in a sustainable production logic. But higher levels of resource efficiency can be achieved only if all actors in the supply chain focus on sustainability goals. For example, the production department should work with only suppliers that have adopted sustainability policies (Carter & Jennings, 2004; Gold, Seuring, & Beske, 2010). From the point of view of human resource management, it has been argued that employees who work in organizations that are concerned with sustainability show greater motivation at work and have a lower turnover rate (Daily & Huang, 2001;

Moon, 2007). This is particularly important if we observe that among the stakeholders that most contribute to the positive performance of companies, based on sustainable or socially responsible practices, are employees (Barreira, F., & Machado, 2019). Therefore it is important to take measures to support employees based on CSR and family-friendly practices (Faria & Machado, 2018).

But we can then ask: How can organizations contribute to sustainability? On one hand, this can be done by using ethical and socially responsible practices (Santiago, Santos, & Carvalho, 2019). On the other hand, innovation—more properly, sustainable innovation—can also be the answer to this question (Kneipp, Gomes, Bichueti, Frizzo, & Perlin, 2019).

The importance of innovation is undeniable, especially from the perspective of competitiveness (Alves, Galina, & Dobelin, 2018). However, understanding the meaning of innovation is somewhat complex, especially in the academic field (Birkinshaw, Hamel, & Mol, 2008), in which research is largely focused on technological innovation. This is because innovation is completely dependent on several mixed actions, new or transformed, for the company to really increase its competitiveness. Therefore to gain a broad understanding of the meaning of innovation, it is necessary to understand it beyond technological change. Organizational innovation is often called administrative innovation (Damanpour, 2014). Organizational innovation can also be seen as a set of company-specific management activities that are new to the company and can be associated with setting goals, motivating employees, coordinating activities, and making decisions, which arise as a result of new dynamics that aim to promote organizational goals (Hollen, Van den Bosch, & Volberda, 2013).

Armbruster, Bikfalvi, Kinkel, and Lay (2008) mentioned the existence of three types of organizational innovation—structural innovation, procedural innovation, and external relations—as a different dimension in which the company's focus can be intraorganizational or interorganizational.

Dyck and Silvestre (2018) report that there is a growing awareness of society innovating to deal with socioenvironmental crises through the adoption of more sustainable lifestyles. Thus a key factor in this context is to implement innovations that promote sustainable development. Sustainable innovation is therefore the creation of something new that improves the three dimensions of sustainable development: social, environmental, and economic. These improvements are not limited to technological changes and may also be related to changes in processes, operational practices, and business models, for example (Szekely & Strebel, 2013). Adams, Jeanrenaud, Bessant, Denyer, and Overy (2016) state that sustainability-oriented innovation is related to changing organizational philosophy and values and to products, processes, or practices that have a specific purpose of creating social and environmental value in addition to profit. Indeed, the adoption of sustainable innovation practices can affect business performance.

In fact, several studies have associated the results of investments in sustainable innovation with business performance (Gunday, Ulusoy, Kilic, & Alpkan, 2011; Lopez-Valeiras, Gomez-Conde, & Naranjo-Gil, 2015). Sustainable innovation thus contributes to business sustainability, as it promotes organization's financial, social, and environmental performance (Aguilera-Caracuel & Ortiz-de-Mandojana, 2013). Lopez-Valeiras et al. (2015) also reinforce that there is a growing interest in the academic, business, and political areas in terms of the relationship between sustainable innovation and business. Gunday et al. (2011) analyzed 184 manufacturing companies in Turkey and determined the positive effects of innovation on business performance issues, such as innovative, productive, market and financial performance. Wagner (2010) also analyzed the relationship between sustainability management and economic performance. He used different measures for social and environmental performance and concluded that the latter has a direct effect on economic performance and the former has only a moderate effect. In fact, sustainable innovation can contribute to the competitive advantage of organizations because of its great importance to stakeholders.

For Schaltegger, Lüdeke-Freund, and Hansen (2016), business activities are responsible for several social and environmental problems, which is why sustainability is so important.

Charter and Clark (2007) argue that sustainable innovation is a subjective concept. However, it is related to entrepreneurship and new technologies, products, services, processes, and social systems. They argue that although the sustainable innovation and eco-innovation are similar concepts, eco-innovation is related only to environmental and economic dimensions, while sustainable innovation is also concerned with ethical and social issues. For Boons, Montalvo, Quist, and Wagner (2013), sustainable innovation goes beyond eco-innovation, including social goals, and refers more clearly to the holistic and long-term process of sustainable development. Siqueira and Pitassi (2016) argue that sustainability-oriented innovation is a concept of eco-innovation, as it integrates three major levels: the macro level, which encompasses government policies and actions that surpass the risks involved in more extreme innovations; the company level, which concerns the development of new business models; and the individual level, based on changes in people's cognition. Hansen, Grosse-Dunker, and Reichwald (2009) note that sustainable innovation integrates sustainability issues and the inclusion of new customers and market segments, adding positive value to the company's global capital. Thus the great challenge for organizations is to innovate through sustainable development.

To assess and structure the effects of sustainable innovations, Hansen et al. (2009) propose a generic model called the Sustainability Innovation Cube, with three dimensions: goal, life cycle, and types of innovation. The goal dimension explores the concept of TBL, distinguishing innovations in the economic, environmental, and social fields. The life cycle dimension is related to the effects of products and technologies at different stages. The types of innovation consider technological innovations in products or processes in business models and Product-Service Systems (PSS). Through the three dimensions of the generic model, Hansen et al. (2009) identified practical implications for the management of sustainable innovation, represented by the following categories: integration of the sustainability criterion, integration of stakeholders, PSS expansion, marketing aimed at sustainable innovation, and sustainability awareness (Kneipp et al., 2019).

Bocken, Short, Rana, and Evans (2014) identified several examples of solutions that can contribute to the innovation of business sustainability and proposed eight assumptions, related to technological, social and organizational dimensions, to describe sustainable mechanisms: improve material and energy efficiency, creation of value from waste, adoption of renewable and natural processes, property replacement for functionality, leadership, spirit of sufficiency, adaptation of companies to society and the environment, and development of joint solutions. These assumptions have the potential to incorporate sustainability into innovations and reduce any risks associated with their implementation. The authors also argue that the main goal is not only to reduce negative social and environmental issues, but also to help reshape business models of sustainability (Bocken, et al., 2014).

Nidumolu, Prahalad, and Rangaswami (2009), when studying sustainable initiatives by large organizations, discovered that success is related to the fact that sustainability is seen as a new line of innovation. Successful companies tend to integrate sustainability with innovation, and they can achieve a competitive advantage by redefining products, services, technologies, procedures, and business models and can reduce costs.

Zee, Fok, and Hartman (2011), in a study of small and large companies in Germany, found that large companies are "greener" than the smaller ones, but small businesses have higher levels of environmental awareness.

Robinson and Stubberud (2013) claim that large companies have "greener" innovation practices because they also have more capital. However, small companies, competing with larger companies, can find in environmental innovation an effective and sustainable mechanism that is more attractive to consumers. Within the scope of this idea, Klewitz and Hansen (2014) analyzed innovation oriented toward sustainability in small and medium-sized companies, and identified products, processes, and practices of organizational innovation and developed an integrated framework of sustainability-oriented innovation for small and medium-sized companies.

Thus we can see that the importance of environmental issues in terms of research and development has been growing because organizations want to promote eco-innovation (Kobayashi, Kato, Maezawa, & Sano, 2011).

But if innovation and sustainability must go hand in hand, the same happens in the case of CSR (Belomonte, 2019; Szutowski & Ratajczak, 2016). Indeed, innovation has become a key to understanding the link between CSR and a company's social and financial performance. Nidumolu et al. (2009) argue that CSR is a fundamental for innovation. Rexhepi, Kurtishi, and Bexheti (2013) argue that CSR and innovation are the foundation of business skills. Furthermore, companies operate in a dynamic environment (Gunday, et al., 2011), and this context forces them to constantly look for new solutions that guarantee the achievement or maintenance of a competitiveness. Therefore innovation is crucial and inevitable for all companies. Also, companies seem to seek socially responsible practices because CSR positively influences a company's performance, improving its reputation (Lou & Bhattacharya, 2006). Therefore it seems essential that companies promote an integration between the two concepts.

Also in this regard, Halme and Laurila (2009) concluded that there are three types of CSR: philanthropic, integration, and innovation. The philanthropic type of CSR is based on employee sponsorship, charity, and volunteerism. The integration type focuses on more responsible business operations. The innovation type promotes the idea of new business models for solving social and environmental problems. Torugsa, O'Donohue, and Hecker (2012) distinguished between reactive and proactive CSR, with proactive CSR being the link to innovation, as it deals with improvements in the company's main activity.

But if understanding these concepts and their interconnection is of paramount importance, it is even more relevant to understand them in the current context, which is marked by a pandemic that has put organizations and all these practices to the test.

6.3 Corporate social responsibility, business ethics, sustainability, and innovation: the impact of the COVID-19 pandemic

At the end of 2019 the COVID-19 pandemic began in Wuhan, China (Zhang, Chen, Zhang, Roy, & Shen, 2020). Over the next months, the pandemic caused a serious economic and social crisis. In fact, the COVID-19 pandemic is the most serious health crisis of this century (Chakraborty & Maity, 2020). Indeed, the COVID-19 crisis has placed huge pressure on the global economy and business activities with significant adverse financial consequences, decreasing countries' gross domestic product and increasing world poverty (Iwuoha, Chiwuike,

Jude-Iwuoha, & Awoke, 2020). Thus the COVID-19 pandemic is a multilevel crisis that has affected individuals, organizations, and countries, jeopardizing the survival of many individuals and the achievement of countless goals (Kovoor-Misra, 2020).

Indeed, May and Mackin (2020) argue that the response to the crisis imposed by COVID-19 must be oriented according to moral, ethical, and socially responsible principles and by a change in corporate governance. Organizational leaders must promote ethical plans and responsible actions. The main objective must be to ensure the health of employees above any business decision. Companies have the critical know-how and resources to create opportunities to help the world and contribute to its recovery.

COVID-19 presents challenges for companies and organizations regarding CSR (Patuelli, Caldarelli, Lattanzi, & Saracco, 2021). Some instances of companies seeking to take advantage of this crisis have been reported. Inevitably, this crisis is putting companies to the test of their commitment to ethical business conduct and to CSR. In some cases, companies seek to make short-term gains, either through fraudulent actions or misconduct or by reducing their investment in CSR. Fortunately, there are many other companies that have chosen the opposite path, reinforcing not only their ethical conduct, but also their socially responsible action, which has become even more evident in the case of organizations that can offer more help and immediate assistance in the fight against the pandemic. There is no doubt that the pandemic offers a wide range of opportunities for companies that take a more aware and insightful approach to CSR. For example, several manufacturing companies transformed their factories to produce masks and personal protective equipment, and in many cases promoted the donation of these products. Several distribution chains provided specific opening hours for older adults and healthcare professionals and promoted the donation of foodstuffs to charitable organizations. Other companies worked in partnership with solidarity organizations to reduce social isolation and loneliness among older people. Therefore the COVID-19 pandemic offers great opportunities for companies to actively engage with their CSR strategies and agendas (He & Harris, 2020).

However, the pandemic has pushed some companies out of the market, so it is important to understand what drives companies to be more ethical and socially responsible, particularly when resources are constrained and their survival is under threat. In fact, we can see that governments around the world have established economic aid packages to alleviate the impending pressure, particularly on the most vulnerable companies, such as small businesses and companies in the most critical sectors (e.g., tourism). In many cases, these measures have encouraged companies to maintain ethical business practices and meet their CSR commitments. There is also evidence that during crises and in a scenario of uncertainty, many leaders stand out at this level. For example, Jack Ma, the cofounder of Alibaba, through his foundation donated coronavirus test kits and other medical supplies to many countries. Therefore this pandemic represents a great context for examining how institutional factors and leadership influence can promote CSR and ethical business conduct in times of crisis (He & Harris, 2020).

But is this the path that most companies will follow? On one hand, the pandemic may discourage companies from investing in CSR, owing to the need to focus on their core operating business for short-term survival. On the other hand, the pandemic can reinforce this investment in response to the growing demands and needs of the market. Therefore if we take a more optimistic view, the COVID-19 pandemic may accelerate the development of postpandemic CSR in the long term as more and more companies come to realize that their survival and sustainable development depend on a delicate balance between profit and their various stakeholders.

Nevertheless, it is unquestionable that the global health crisis caused by this pandemic has tremendously delayed the international community's progress toward sustainability (Lee, Kang, & Kim, 2020). Several studies have been carried out in terms of the impact of the pandemic on sustainability (Ranjbari, et al., 2021; Leal Filho, Brandli, Lange Salvia, Rayman-Bacchus, & Platje, 2020).

Barbier and Burgess (2020) argue that the adverse impacts of COVID-19 could compromise our ability to achieve several of the 17 U.N. Sustainable Development Goals (SDG) within the 2030 Agenda for Sustainable Development. Furthermore, Leal Filho et al. (2020) identified COVID-19 as a significant threat to the implementation of sustainable development, reducing the priority of SDG. Therefore while significant global efforts are being made to control this pandemic, sustainability in the post-COVID-19 era must not be devalued (Lambert et al., 2020). In fact, sustainability and compliance with the SDG are even more critical now than ever (Leal Filho et al., 2020).

It is also important to note that, given the way in which the crisis instigated by COVID-19 is spreading, its impact differs from country to country. The different ways in which organizations apply certain CSR practices also varies from country to country, depending on certain variables that influence these practices, such as culture (Magalhães, 2011). This is independent of the pandemic. The same occurs in the case of sustainability, whose practices are strongly influenced by normative and cultural contexts (Horak et al., 2018). At the same time, we see that the impact of this pandemic also varies socially, as there are social groups that are more vulnerable than others (Tanna, Conti, & Silva, 2021). Population aging is also be a variable that must be considered in assessing the different impacts that the pandemic ha had (Abbas, et al., 2021). This makes it difficult to identify a viable solution for all and for policymakers to identify quick and short-term mechanisms to manage the consequences of this crisis. Therefore governments and the business community need to assess both short-term responses and long-term value maximization.

A curiosity concerns the fact that innovative start-ups are considered better prepared than many other types of companies to deal with the crisis caused by COVID-19 (Kuckertz et al., 2020). Studying how these types of start-ups handle crises like this one will be critical (Williams & Vorley, 2015). The same is true of resilience (Doern, Williams, & Vorley, 2019). This occurs because most published studies of crisis management were carried out before this pandemic.

Also, the theme of the impact of COVID-19 on sustainability and CSR was explored by some academics in Asia. Indeed, there are many lessons that can be learned from this pandemic crisis, and the resilience of some organizations has led to sustainable practices that promote organizational development, even though the focus of action has changed in some cases (Schaltegger, 2021). Zhang et al. (2020) carried out an empirical study and found that the sustainability priorities of Chinese companies shifted toward the social dimension. Nevertheless, companies have been giving more importance to economic sustainability in the pandemic period, and environmental sustainability has lost priority. These results were identical to those of another study carried out in China, which concluded that the environmental dimension was given a lower priority than the social one (Zhang, Lu, Morse, & Liu, 2021). Liu et al. (2020) focused their research on China, South Korea, and Singapore and examined the economic implications of COVID-19 from the theoretical foundations of entrepreneurship, agility, and resilience. They concluded that COVID-19 resulted in a dissociation of China's supply chains and the relocation of strategic manufacturing operations out of China, and they argued that it is very important that governments and industries work together to minimize disruption to global supply chains in the postpandemic period. Haak-Saheem (2020) analyzed companies in Dubai and discussed the impact of COVID-19 on talent management. The author concluded that the post-COVID-19 job market in Dubai will face stronger competition for jobs and talent; greater flexibility of work, space, and time; active support to help individuals find jobs; and greater investment in talent attraction. Yoshikawa and Chua (2020) analyzed the relationship between shareholder activism in Japan and COVID-19, and Gao and Ren (2020) discussed the challenges facing global value chains because of COVID-19 and how the current dependence of Asian countries can be reformulated by this pandemic.

Other authors from around the world have also tried to explore the impact of COVID on business, sustainability, and CSR out of Asia. Barbier and Burgess (2020), for example, identified three accessible and innovative policy mechanisms for developing countries to achieve sustainable development after COVID-19. He and Harris (2020) analyzed how COVID-19 could influence the development of CSR and concluded that the pandemic offered an opportunity for companies to shift toward a more genuine and authentic CSR. Palma-Ruiz, Castillo-Apraiz, and Gómez Martínez (2020) investigated the investment strategies of Spanish trading companies during the pandemic and found that socially responsible actions could be a competitive advantage for Spanish companies. Barreiro-Gen, Lozano, and Zafar (2020) explored changes in corporate sustainability priorities because of COVID-19, analyzing 653 organizations worldwide. The authors found that the priority is now the social dimension for most organizations and not the environmental one, which is in line with the studies mentioned previously. Garcia-Sanchez & Garcia-Sanchez (2020) investigated the CSR of large Spanish companies during COVID-19 and determined the importance of economic interests to most of them.

Indeed, the structural transition that many organizations are experiencing as a result of the pandemic makes it urgent to rethink business models, which should increasingly have an innovative and sustainable action at their origin. We cannot forget that the concept of degrowth was proposed as an important approach for the economy and society to become more sustainable (Sekulova, Kallis, Rodríguez-Labajos, & Schneider, 2013).

From an environmental point of view, even given that, according to some studies, the environmental aspect is losing priority to the social aspect, there was a 5.8% worldwide reduction in CO_2 emissions during the first quarter of 2020, ranging from 10.3% for China to 4.2% for the United States, compared to 2019 (Liu et al., 2020). Le Quéré et al. (2020) also indicate that carbon emissions in countries decreased, on average, by 26%. In other words, the pandemic ended up having positive effects at the environmental level, contrary to what happened in social terms, which may be the main reason why many countries and organizations focused more on investments of a social nature.

However, COVID-19 also had negative effects on transforming the sustainability of industries. The pandemic affected global low-carbon energy supply chains because closed borders led to component scarcity, leading to lower production of wind turbines, solar panels, and electric vehicles, among others (Goldthau & Hughes, 2020). Thus the energy transition requires an intelligent transition of business sustainability (Köhler et al., 2019; Schaltegger et al., 2016), which can be achieved at the expense of more innovative models and processes.

In fact, several companies have made interesting profits through innovations that they have incorporated into their business models. When organizations innovate, they add new value to a product, service, process, management system, or market (Crossan & Apaydin, 2010). Organizational innovation is complex and integrates the following components: innovation drivers; the process of idealizing, developing, and implementing the innovation; and the results of innovation (Garud et al., 2013). Innovation drivers are the triggers that initiate innovation in the organization and can be external or internal. External drivers can be new regulations, new markets, and even a crisis such as COVID-19; internal motivators can be new knowledge and resources (Zouaghi, Sanchez, & Martinez, 2018).

A crisis can motivate and be a driver of the organization for innovation, because of the attributes of the situation. Furthermore, several stakeholders are affected by the crisis and are involved in its resolution. Thus it is important to note that a crisis also brings opportunities for organizations to innovate, learn, strengthen relationships with stakeholders, improve the organizational climate, and grow (Kovoor-Misra, 2020). These attributes also apply to the COVID-19 pandemic.

Indeed, during the pandemic, organizations have faced several challenges, but they also can learn and innovate to solve problems. This innovation has an impact on the progress of companies after a crisis (Nemlioglu & Mallick, 2020), especially when organizations can improvise, innovate, and solve problems to resolve the crisis (Kovoor-Misra, 2020). Thus innovation serves as a mean for survival and a way to overcome challenges for an organization in crisis (Glodzinski & Marciniak, 2016; Klodane & Zvaigzne, 2017; Vergne & Depeyre, 2016).

However, it is useful to examine the extent to which organizations can be innovative and resilient during a crisis, being influenced by their resources, productive system capabilities, and relationships with stakeholders and the adaptability of their products or services to emerging contexts (Kovoor-Misra, 2020). What has been observed is that a good basis in terms of leadership, technology, human resources, infrastructure, innovation capacity, and positive relationships with the several stakeholders allows an organization to respond quickly to a crisis (Kovoor-Misra, 2020).

At this point, it is worth asking: To what extent can an organization's products or services be innovated within the constraints of social interactions brought about by the pandemic? There are studies that show that in some cases, the pandemic accelerated innovation, also promoting ethical and socially responsible practices (Gopalakrishnan & Kovoor-Misra, 2021). In the case of hospitals, for example, innovation was necessary, as these organizations had to create new structures to provide health services in a suddenly arising new context. Some hospitals have partnered with technology companies, which have given them tools to innovate and provide remote services, such as virtual visits, digital patient screening, remote monitoring to identify mild COVID-19 symptoms, cameras to measure body temperature, and digital platforms to analyze the mental health of employees. It is also worth mentioning pharmaceutical companies, which had to adapt quickly in terms of administering vaccines, testing for COVID-19 infection, and supplying other products and auxiliary medical equipment (Lee & Trimi, 2021). Another example to highlight is that of airlines, many of which started using their planes to transport goods, food, and medicines. Some airlines created process-level innovations, such as new procedures to promote distance between passengers and increasing crew levels to provide additional sanitation measures. Many other companies have innovated their processes by partnering with healthcare institutions to provide tests and healthcare for their employees. Companies in the automotive sector started to produce fans, for example, expanding their market and, at the same time, adopting an ethical and socially responsible practice. Another example is passenger transport companies such as Uber, which saw demand for their services drop drastically and had to adapt to this new scenario. Uber renewed its platform and created a new application called Work Hub so that its drivers could find other jobs at Uber Eats (food delivery) or Uber Freight (goods delivery). In streaming companies, for example, the pandemic created conditions that exponentially increased demand for their products and the emergence of new markets. Through this, several companies have developed platforms to promote remote social contact and distance entertainment. Companies promoting videoconferencing services saw their services expand and had to create more extensive forms to adapt to the reality of their provision, owing to the increase in remote work and telemedicine, among others. The same happened with companies in the field of artificial intelligence. Several technology companies in the computer sector improved their network capacity and download speed (Gopalakrishnan & Kovoor-Misra, 2021). Reference should also be made to educational institutions, which had to develop and implement, in record time, distance learning platforms (Lee & Lee, 2020). It should also be noted that the urgency of making certain products and services available has resulted in many partnerships between scientists, private foundations, pharmaceutical companies, and other organizations, such as universities, through their research centers (Gopalakrishnan & Kovoor-Misra, 2021).

In short, all over the world, we have been witnessing new ways of acting by companies, with ethical and CSR practices adapted to this new context, marked by a greater concern with innovation, as the organizations have had to readjust and even reinvent. It is only on the basis of sustainable strategies that we can assume that these practices will have an effect today and tomorrow, especially after organizations have realized that they are exposed to enormous uncertainty and instability and that if they are not quick in their response to crises like the one in which we currently live, their own survival may be jeopardized. Finally, we must not forget that the study of these themes, namely, sustainability, must consider the various interactions that occur at this level, especially from the social and environmental point of view, so that none of the relevant scenarios are neglected (Selin, 2021).

6.4 Conclusion

We can conclude that the pandemic has demonstrated that organizations need to innovate in a sustainable way, to be able to face adverse scenarios like the one we are currently experiencing. If the pandemic was a threat for some organizations, for others it was an opportunity to innovate and adopt more sustainable and therefore more efficient practices.

At the outset, what is at stake is not an understanding of the concepts analyzed here, but their application. Many companies associate CSR with a cost and not an investment, and many do not have the know-how or capital to invest in innovation, especially with a sustainable focus. Others see sustainability as a concept with long-term effects, which reduces the concern around it, especially in the light of more immediate mentalities. All of this can promote less ethical behaviors in organizations, especially if the survival of the business is at stake.

We also found that during the COVID-19 pandemic, many organizations were more resilient, due to their adaptability, innovation, and collaborative spirit. Many companies took advantage of the recovery packages to promote sustainable management practices, not only because they realized that this is the correct way to manage resources, but also because it is the way to better promote the business in a long-term way. In other words, it is not just about doing what is right, but also what is necessary for organizational competitiveness.

Concepts such as a circular economy, transposed to manufacturing (circular manufacturing), can and should be promoted by government action through support and awareness-raising practices, drawing attention to the positive effect they can have on a business, which must be integrative, inclusive, and sustainable, promoting greater organizational efficiency.

The pandemic exposed the fragility of the worldwide industrial production system, causing scarcity of many products and services. This should lead us to think about the important role that manufacturing has and the importance of its restructuring. Concepts such as just-in-time must now be revised, as should the current globalized production structure, which makes organizations very interdependent and, in some processes, not autonomous. The transfer of knowledge to manufacturing is essential for its innovation and sustainable management. This is why some organizations in certain countries, such as China, were more able to produce personal protective equipment than others. Indeed, the ability to integrate the production process, with access to raw materials and know-how, are essential for sustainable innovation. Likewise, intraorganizational and interorganizational collaboration can add value to companies and increase their ability to respond to crises.

If thinking globally is important to create a collective conscience in global terms and promote synergies between countries and organizations, acting locally is essential for establishing networks that are capable of responding quickly to emerging problems, promoting sustainability through support of local suppliers and manufacturing in the places of origin of raw materials, which reduces costs with logistics and storage, for example. Eco-industrial parks, for instance, can be a sustainable and more efficient way of managing the entire value chain of the production system.

Indeed, COVID-19 brought organizations the opportunity to rethink their manufacturing in a more ecological, circular, and integrated way, taking local and regional needs fully into account. Although global strategies should not be overlooked or made irrelevant, local strategies should be promoted to increase synergies between organizations from several sectors, such as companies, universities, hospitals, and nongovernmental organizations, among others. That is what the pandemic showed and taught us: that joining efforts around a common goal can be more productive and even a matter of survival.

Never before has the CSR stakeholder theory made more sense. Indeed, all parties that are directly and indirectly interested and affected by the business of organizations must be an active part of this change that is claimed and that must be based on innovative sustainable practice and, ethically and socially responsible—that is, sustainable—management. Thus economic recovery packages made possible by governments around the world must prioritize these practices. Perhaps this is the biggest and best opportunity for companies to promote a type of management that is intended to be increasingly sustainable.

References

Abbas, H. S. M., et al. (2021). Impact of COVID-19 pandemic on sustainability determinants: A global trend. *Helyon, 7*, e05912.

Adams, R., Jeanrenaud, S., Bessant, J., Denyer, D., & Overy, P. (2016). Sustainability-oriented innovation: A systematic review. *International Journal of Management Reviews, 18*(2), 180–205.

Aguilera-Caracuel, J., & Ortiz-de-Mandojana, N. (2013). Green innovation and financial performance: An institutional approach. *Organization & Environment, 26*(4), 365–385.

Alves, M. F. R., Galina, S. V. R., & Dobelin, S. (2018). Literature on organizational innovation: Past and future. *Innovation and Management Review, 15*(1), 2–19.

Armbruster, H., Bikfalvi, A., Kinkel, S., & Lay, G. (2008). Organizational innovation: The challenge of measuring non-technical innovation in large-scale surveys. *Technovation, 28*(10), 644–657.

Babiak, K., & Trendafilova, S. (2011). CSR and environmental responsibility: Motives and pressures to adopt green management practices. *Corporate Social Responsibility and Environmental Management, 18*(1), 11–24.

Barbier, E. B., & Burgess, J. C. (2020). Sustainability and development after COVID-19. *World Development, 135*, 105082.

Barreira, A., F., & Machado, C. F. (2019). Transversal competences: A true and effective support to achieve greater organizational sustainability. In C. Machado (Ed.), *Sustainable human resource management: Policies and practices*. River Publishers.

Barreiro-Gen, M., Lozano, R., & Zafar, A. (2020). Changes in sustainability priorities in organisations due to the COVID-19 outbreak: Averting environmental rebound effects on society. *Sustainability, 12*, 5031.

Barry, N. (2002). The stakeholder concept of corporate control is illogical and impractical. *The Independent Review, 6*(4), 541–554.

Belomonte, L. J. (2019). The sustainable approach to corporate social responsibility: A global analysis and future trends. *Sustainability, 11*(19), 5382.

Birkinshaw, J., Hamel, G., & Mol, M. J. (2008). Management innovation. *Academy of Management Review, 33*(4), 825–845.

Black, L. D., & Härtel, C. E. J. (2004). The five capabilities of socially responsible companies. *Journal of Public Affairs, 4*(2), 125–144.

Bocken, N. M. P., Short, S. W., Rana, P., & Evans, S. (2014). A literature and practice review to develop sustainable business model archetypes. *Journal of Cleaner Production, 65*(1), 42–56.

Boons, F., Montalvo, C., Quist, J., & Wagner, M. (2013). Sustainable innovation, business models and economic performance: An overview. *Journal of Cleaner Production, 45*(1), 1–8.

Bowen, H. (1953). *Social responsibility of the businessman*. New York: Harper and Row.

Carroll, A. B., & Schwartz, M. (2003). Corporate social responsibility: A three-domain approach. *Business Ethics Quarterly, 13*(4), 503–530.

Carroll, A. B. (1979). A three-dimensional conceptual model of corporate performance. *Academy of Management Review, 4*(4), 497–505.

Carroll, A. B. (1991). The pyramid of corporate social responsibility: Toward the moral management of corporate stakeholders. *Business Horizons, 34*(4), 39–48.

Carroll, A. B. (1998). The four faces of corporate citizenship. *Business and Society, 100*(1), 1–7.

Carter, C. R., & Jennings, M. M. (2004). The role of purchasing in corporate social responsibility. A structural equation analysis. *Journal of Business Logistics, 25*(1), 145–186.

Chakraborty, I., & Maity, P. (2020). COVID-19 outbreak: Migration, effects on society, global environment and prevention. *Science of the Total Environment, 728*, 138882.

Charter, M., & Clark, T. (2007). *Sustainable innovation – key conclusions from sustainable innovation conferences 2003–2006 organised by the centre for sustainable design. The centre for sustainable design*. Farnham, UK: University College for the Creative Arts.

Chen, Y. (2010). The drivers of green brand equity: Green brand image, green satisfaction and green trust. *Journal of Business Ethics, 93*(2), 307–319.

Cohen, J., Pant, L., & Sharp, D. (1993). A validation and extension of a multidimensional ethics scale. *Journal of Business Ethics, 12*(1), 13–26.

Crossan, M. M., & Apaydin, M. (2010). A multi-dimensional framework of organizational innovation: A systematic review of the literature. *Journal of Management Studies, 47*(6), 1154–1191.

Daily, B. F., & Huang, S. (2001). Achieving sustainability through attention to human resource factors in environmental management. *International Journal of Operations & Production Management, 21*(12), 1539–1552.

Damanpour, F. (2014). Footnotes to research on management innovation. *Organization Studies, 35*(9), 1265–1285.

Dangelico, R. M., & Pujari, D. (2010). Mainstreaming green product innovation: Why and how companies integrate environmental sustainability. *Journal of Business Ethics, 95*(3), 471–486.

Ditlev-Simonsen, C. D., & Midttun, A. (2011). What motivates managers to pursue corporate responsibility? A survey among key stakeholders. *Corporate Social Responsibility and Environmental Management, 18*(1), 25–38.

Doern, R., Williams, N., & Vorley, T. (2019). Special issue on entrepreneurship and crises: Business as usual? An introduction and review of the literature. *Entrepreneurship and Regional Development, 31*(5–6), 400–412.

Dunphy, D., Griffiths, A., & Benn, S. (2007). *Organizational change for corporate sustainability. A guide for leaders and change agents of the future* (2nd ed.). London: Routledge.

Dyck, B., & Silvestre, D. S. (2018). Enhancing socio-ecological value creation through sustainable innovation 2.0: Moving away from maximizing financial value capture. *Journal of Cleaner Production, 171*(1), 1593–1604.

Elkington, J. (1998). *Cannibals with forks: The triple bottom line of 21st century business*. Stony Creek, CT: New Society Publishers.

Enderle, G. e, & Tavis, L. A. (1998). A balanced concept of the firm and the measurement of its long-term planning and performance. *Journal of Business Ethics, 17*(11), 1129–1144.

Evan, W. M., & Freeman, R. E. (1993). A stakeholder theory of the modern corporation: Kantian capitalism. In T. L. Beauchamp, & N. E. Bowie (Eds.), *Ethical theory and business* (pp. 75–84). Prentice-Hall.

Faria, A., & Machado, C. F. (2018). Work-family conciliation policies: Answering to corporate social responsibility – a case study. In C. F. Machado, & P. J. Davim (Eds.), *Corporate social responsibility in management and engineering*. River Publishers.

Frederick, W. (1986). Toward CSR3 – why ethical analysis is indispensable and unavoidable in corporate affairs. *California Management Review, 28*(2), 126–141.

Frederick, W. (1994). From CSR1 to CSR2. *Business and Society, 33*(2), 150–164.

Frederick, W. (1998). Moving to CSR4: What to pack for the trip. *Business and Society, 37*(1), 40–59.

Freeman, E. (1984). *Strategic management: A stakeholder approach*. Boston, MA: Pitman Publishing.

Freeman, E., & Reed, D. L. (1983). Stockholders and stakeholders: A new perspective on corporate governance. *California Management Review, 25*(3), 88–106.

Friedman, M. (1962). *Capitalism and freedom*. The University Chicago Press.

Friedman, M. (1970). The social responsibility of business is to increase its profits. *New York Times Magazine, 13*, 122–126.

Gao, H., & Ren, M. (2020). Overreliance on China and dynamic balancing in the shift of global value chains in response to global pandemic COVID-19: An Australian and New Zealand perspective. *Asian Business & Management, 19*, 306–310.

Garcia-Sanchez, I., & Garcia-Sanchez, A. (2020). Corporate social responsibility during COVID-19 pandemic. *Journal of Open Innovation: Technology, Market, and Complexity, 6*(4), 1–21.

Garriga, E., & Melé, D. (2004). Corporate social responsibility theories: Mapping the territory. *Journal of Business Ethics, 53*(12), 51–71.

Garud, R., et al. (2013). Perspectives on innovation processes. *The Academy of Management Annals, 7*(1), 775–819.

Gawel, A. (2012). Entrepreneurship and sustainability: Do they have anything in common? *Poznan University of Economic Review, 12*(1), 5–16.

Geissdoerfer, M., Savaget, P., Bocken, N., & Hultink, E. (2017). The circular economy – a new sustainability paradigm? *Journal of Cleaner Production, 143*(1), 757–768.

Gimenez, C., Sierra, V., & Rodon, J. (2012). Sustainable operations: Their impact on the triple bottom line. *International Journal of Production Economics, 140*(1), 149–159.

Glodzinski, E., & Marciniak, S. (2016). Organizational innovations in crisis management of project-based enterprises. *Economics and Business, 28*(1), 26–32.

Gold, S., Seuring, S., & Beske, P. (2010). Sustainable supply chain management and inter-organizational resources: A literature review. *Corporate Social Responsibility and Environmental Management, 17*(4), 230–245.
Goldthau, A., & Hughes, L. (2020). Protect global supply chains for low-carbon technologies. *Nature, 585*, 28–30.
Gopalakrishnan, S., & Kovoor-Misra, S. (2021). Understanding the impact of the COVID-19 pandemic through the lens of innovation. *BRQ Business Research Quarterly, 24*(3), 224–232.
Gunday, G., Ulusoy, G., Kilic, K., & Alpkan, L. (2011). Effects of innovation types on firm performance. *International Journal of Production Economics, 133*(2), 662–676.
Gupta, S., & Kumar, V. (2013). Sustainability as corporate culture of a brand for superior performance. *Journal of World Business, 48*(3), 311–320.
Haak-Saheem, W. (2020). Talent management in COVID-19 crisis – how Dubai manages and sustains its global talent pool. *Asian Business and Management, 19*(3), 298–301.
Hahn, T., Figge, F., Aragón-Correa, J. A., & Sharma, S. (2015). Advancing research on corporate sustainability – off to pastures new or back to the roots? *Business & Society, 56*(2), 155–185.
Halbusi, H., Ismail, M. N., & Omar, S. B. (2021). Ethical leadership and employee ethical behaviour: Exploring dual-mediation paths of ethical climate and organisational justice: Empirical study on Iraqi organisations. *International Journal of Business Governance and Ethics, 15*(3), 303–325.
Halme, M., & Laurila, J. (2009). Philanthropy, integration or innovation? Exploring the financial and societal outcomes of different types of corporate responsibility. *Journal of Business Ethics, 84*, 325–339.
Hansen, E., Grosse-Dunker, F., & Reichwald, R. (2009). Sustainability innovation cube: A framework to evaluate sustainability-oriented innovations. *International Journal of Innovation Management, 13*(4), 683–713.
Hansen, R. S. (1992). A multidimensional scale for measuring business ethichs: A purification and refinement. *Journal of Business Ethics, 11*(7), 523–534.
He, H., & Harris, L. (2020). The impact of COVID-19 pandemic on corporate social responsibility and marketing philosophy. *Journal of Business Research, 116*, 176–182.
Hemphill, T. (2004). Monitoring global corporate citizenship: Industry self-regulation at a crossroads. *Journal of Corporate Citizenship, 14*, 81–95.
Hollen, R. M. A., Van den Bosch, F. A. J., & Volberda, H. W. (2013). The role of management innovation in enabling technological process innovation: An inter-organizational perspective. *European Management Review, 10*(1), 35–50.
Horak, S., Arya, B., & Ismail, K. M. (2018). Organizational sustainability determinants in different cultural settings: A conceptual framework. *Business Strategy and the Environment, 27*(4), 528–546.
Iwuoha, J. C., Chiwuike, U., Jude-Iwuoha, A. U., & Awoke, F. C. (2020). The politics of inequality and social Vices Nexus in Nigeria. *International Journal of Business Education and Management Studies, 5*(1), 23–34.
Johansen, T. S., & Nielsen, A. E. (2012). CSR in corporate self-storying—Legitimacy as a question of differentiation and conformity. *Corporate Communications: An International Journal, 17*(4), 434–448.
Klewitz, J., & Hansen, E. G. (2014). Sustainability-oriented innovation of SMEs: A systematic review. *Journal of Cleaner Production, 65*(1), 57–75.
Klodane, A., & Zvaigzne, A. (2017). Theoretical aspects of innovation in crisis management. *Latgale National Economy Research, 9*, 2747.
Kneipp, J. M., Gomes, C. M., Bichueti, R. S., Frizzo, K., & Perlin, A. P. (2019). Sustainable innovation practices and their relationship with the performance of industrial companies. *Revista de Gestão, 26*(2), 94–111.

Kobayashi, H., Kato, M., Maezawa, Y., & Sano, K. (2011). An R&D management framework for ecotechnology. *Sustainability, 3*(8), 1282–1301.

Köhler, J., et al. (2019). An agenda for sustainability transitions research: State of the art and future directions. *Environmental Innovation and Societal Transitions, 31*, 1–32.

Kok, P., Van der Weile, T., McKenna, R., & Brown, A. (2001). A corporate social responsibility audit within a quality management framework. *Journal of Business Ethics, 31*(4), 285–297.

Kovoor-Misra, S. (2020). *Crisis management: Resilience & change*. Sage.

Kuckertz, A., et al. (2020). Startups in times of crisis – a rapid response to the COVID-19 pandemic. *Journal of Business Venturing Insights, 13*(C).

Lambert, H., Gupte, J., Fletcher, H., Hammond, L., Lowe, N., Pelling, M., ... Shanks, K. (2020). COVID-19 as a global challenge: Towards an inclusive and sustainable future. *The Lancet Planetary Health, 4*, e312ee314.

Le Quéré, C., et al. (2020). Temporary reduction in daily global CO_2 emissions during the COVID-19 forced confinement. *Nature Climate Change, 10*(7), 647–653.

Leal Filho, W., Brandli, L. L., Lange Salvia, A., Rayman-Bacchus, L., & Platje, J. (2020). COVID-19 and the UN sustainable development goals: Threat to solidarity or an opportunity? *Sustainability, 12*, 5343.

Lee, D., Kang, J., & Kim, K. (2020). Global collaboration research strategies for sustainability in the post COVID-19 era: Analyzing virology-related national-funded projects. *Sustainability, 12*, 6561.

Lee, S. M., & Lee, D. (2020). "Untact": A new customer service strategy in the digital age. *Service Business, 14*(1), 1–22.

Lee, S. M., & Trimi, S. (2021). Convergence innovation in a digital age and in the COVID-19 pandemic crisis. *Journal of Business Research, 123*, 14–22.

Levitt, T. (1958). The dangers of social responsibility. *Harvard Business Review, 36*(5), 41–50.

Liu, Z. C., et al. (2020). COVID-19 causes record decline in global CO_2 emissions. UC Berkeley.

Lopez-Valeiras, E., Gomez-Conde, J., & Naranjo-Gil, D. (2015). Sustainable innovation, management accounting and control systems, and international performance. *Sustainability, 7*(3), 3479–3492.

Lou, X., & Bhattacharya, C. B. (2006). Corporate social responsibility, customer satisfaction, and market value. *Journal of Marketing, 70*(4), 1–18.

Magalhães, C. (2011). Responsabilidade social e regulada no sector metalomecânico brasileiro e português: um estudo comparativo e cultural. In N. C. Maria Alice, et al. (Eds.), *Responsabilidade social: Uma visão ibero-americana* (pp. 81–109). Almedina.

Maignan, I. e, & Ferrell, O. C. (2000). Measuring corporate citizenship in two countries: The case of the United States and France. *Journal of Business Ethics, 23*(3), 283–297.

Maignan, I. e, & Ferrell, O. C. (2001a). Antecedents and benefits of corporate citizenship: An investigation of French businesses. *Journal of Business Research, 51*(1), 37–51.

Maignan, I. e, & Ferrell, O. C. (2001b). Corporate citizenship as a marketing instrument: Concepts, evidence and research directions. *European Journal of Marketing, 35*(3–4), 457–484.

Marrewijk, M. V. (2003). Concepts and definitions of CSR and corporate sustainability: Between agency and communion. *Journal of Business Ethics, 44*(2–3), 95–105.

Matten, D., Crane, A. e, & Chapple, W. (2003). Behind the mask: Revealing the true face of corporate citizenship. *Journal of Business Ethics, 45*(1/2), 109–120.

May, D., & Mackin, C. (2020). Responding to the COVID-19 crisis: Public policy and corporate governance considerations. *Challenge, 63*(4), 234–238.

McWilliams, A., & Siegel, D. (2001). Corporate social responsibility: A theory of the firm perspective. *Academy of Management Review, 26*(1), 117–127.

McWilliams, A., Parhankangas, A., Coupet, J., Welch, E., & Barnum, D. T. (2016). Strategic decision making for the triple bottom line. *Business Strategy and the Environment, 25*(3), 193–204.

Mintzberg, H., Simons, R. e, & Kunal, B. (2002). Beyond selfishness. *MIT Sloan Management Review, 44*(1), 67–74.

Moon, J. (2007). The contribution of corporate social responsibility to sustainable development. *Sustainable Development, 15*(5), 296–306.

Nemlioglu, I., & Mallick, S. K. (2020). Do innovation-intensive firms mitigate their valuation uncertainty during bad times? *Journal of Economic Behavior and Organization, 177*, 913–940.

Nidumolu, R., Prahalad, C. K., & Rangaswami, M. R. (2009). Why sustainability is now the key driver of innovation. *Harvard Business Review, 87*(9), 56–64.

Palma-Ruiz, J. M., Castillo-Apraiz, J., & Gómez Martínez, R. (2020). Socially responsible investing as a competitive strategy for trading companies in times of upheaval amid COVID-19: Evidence from Spain. *International Journal of Financial Studies, 8*(3), 41.

Patuelli, A., Caldarelli, G., Lattanzi, N., & Saracco, F. (2021). Firms challenges and social responsibilities during COVID-19: A Twitter analysis. *PLoS One, 16*(7), e0254748.

Paula, K., & Parra, C. M. (2021). Corporate social responsibility in international business literature: Results from text data mining of the *Journal of International Business Studies. International Journal of Corporate Social Responsibility, 6*(12), 1–14.

Peterson, D. K. (2004). The relationship between perceptions of corporate citizenship and organizational commitment. *Business and Society, 43*(3), 296–319.

Phillips, R., et al. (2003). What stakeholder theory is not. *Business Ethics Quarterly, 13*(4), 479–502.

Ranjbari, M., et al. (2021). Three pillars of sustainability in the wake of COVID-19: A systematic review and future research agenda for sustainable development. *Journal of Cleaner Production, 297*, 1–63.

Ranjbari, M., Morales-Alonso, G., Shams Esfandabadi, Z., & Carrasco-Gallego, R. (2019). Sustainability and the sharing economy: Modelling the interconnections. *Dirección y Organización, 68*, 33–40.

Reidenbach, R. E., & Robin, D. P. (1988). Some initial steps towards improving the measurement of ethical evaluations of marketing activities. *Journal of Business Ethics, 7*, 871–879.

Reidenbach, R. E., & Robin, D. P. (1990). Toward the development of a multidimensional scale for improving evaluations of business ethics. *Journal of Business Ethics, 9*, 639–653.

Rexhepi, G., Kurtishi, S., & Bexheti, G. (2013). Corporate social responsibility (CSR) and innovation the drivers of businessgrowth? *Procedia – Social and Behavioral Sciences, 75*, 532–541.

Rhenman, E., & Stymn, B. (1965). *Företagsledning i en Föränderlig Värld [Corporate management in a changing world]*. Aldus/Bonniers.

Robinson, S., & Stubberud, H. A. (2013). Green innovation in Germany: A comparison by business size. *Journal of International Business Research, 12*(1), 47–56.

Salehi, M., Saeidinia, M., & Aghaei, M. (2012). Business ethics. *International Journal of Scientific and Research Publications, 2*(1), 1–4.

Santiago, A., Santos, M., & Carvalho, A. (2019). Organizational sustainability: From theory to practice. *International Journal of Development Research, 9*(5), 27944–27951.

Schaltegger, S. (2021). Sustainability learnings from the COVID-19 crisis. Opportunities for resilient industry and business development. *Sustainability Accounting, Management and Policy Journal, 12*(5), 889–897.

Schaltegger, S., Lüdeke-Freund, F., & Hansen, E. G. (2016). Business models for sustainability: A co-evolutionary analysis of sustainable entrepreneurship, innovation, and transformation. *Organization & Environment, 29*(3), 264–289.

Schawartz, M. (2002). A code of ethics for corporate code of of ethics. *Journal of Business Ethics, 41*(1–2), 27–43.

Scott, R. W. (2014). *Institutions and organizations: Ideas, interests, and identities*. Thousand Oaks, CA: Sage.
Sekulova, F., Kallis, G., Rodríguez-Labajos, B., & Schneider, F. (2013). Degrowth: From theory to practice. *Journal of Cleaner Production, 38*, 1–6.
Selin, N. E. (2021). Lessons from a pandemic for systems-oriented sustainability research. *Science Advances, 7*(22), 1–13.
Siqueira, R. P., & Pitassi, C. (2016). Sustainability-oriented innovations: Can mindfulness make a difference? *Journal of Cleaner Production, 139*, 1181–1190.
Stoney, C., & Winstanley, D. (2001). Stakeholding: Confusion or utopia? Mapping the conceptual terrain. *Journal of Management Studies, 38*(5), 603–626.
Sundman, P. (2000). The good manager: A moral manager. *Journal of Business Ethics, 27*(3), 247–254.
Szekely, F., & Strebel, H. (2013). Incremental, radical and game changing: Strategic innovation for sustainability. *Corporate Governance, 13*(5), 467–481.
Szutowski, D., & Ratajczak, P. (2016). The relation between CSR and innovation model approach. *Journal of Entrepreneurship, Management and Innovation, 12*(2), 77–94.
Tanna, A., Conti, D. M., & Silva, L. F. (2021). Business ethics and corporate social responsibility in times of COVID 19. *Journal on Innovation and Sustainability, 12*(3), 4–11.
Torugsa, N., O'Donohue, W., & Hecker, R. (2012). Capabilities, proactive CSR and financial performance in SMEs: Empirical evidence from an Australian manufacturing industry sector. *Journal of Business Ethics, 109*(4), 483–500.
Vergne, J. P., & Depeyre, C. (2016). A fuzzy-set analysis of the role of cognition and capabilities in U.S. defense firms' responses to 9/11/. *Academy of Management Journal, 59*(5), 1653–1680.
Vos, J. (2003). Corporate social responsibility and the identification of stakeholders. *Corporate Social Responsibility and Environmental Management, 10*, 141–152.
Waddock, S. e, & Smith, N. (2000). Relationships: The real challenge of corporate global citizenship. *Business and Society Review, 105*(1), 47–62.
Wagner, M. (2010). The role of corporate sustainability performance for economic performance: A firm level analysis of moderation effects. *Ecological Economics, 69*(7), 1553–1560.
Warhurst, A. (2001). Corporate citizenship and corporate social investment. *Journal of Corporate Citizenship, 1*, 57–73.
Williams, N., & Vorley, T. (2015). Institutional asymmetry: How formal and informal institutions affect entrepreneurship in Bulgaria. *International Small Business Journal, 33*(8), 840–861.
Wilson, M. (2003). Corporate sustainability: What is it and where does it come from? *Ivey Business Journal, 67*(6), 1–5.
Windolph, S. E., Harms, D., & Schaltegger, S. (2014). Motivations for corporate sustainability management: Contrasting survey results and implementation. *Corporate Social Responsibility and Environmental Management, 25*(1), 272–285.
Wood, J. D., & Logsdon, J. M. (2002). Business citizenship: From domestic to global level of analysis. *Business Ethics Quarterly, 12*(2), 155–187.
Yoshikawa, T., & Chua, G. (2020). COVID-19 and Japanese shareholder activism: Brief respite for Japan's self-healing concrete. *Asian Business and Management, 19*(3), 302–305.
Zee, S. M. L., Fok, L. Y., & Hartman, S. J. (2011). Exploring the relationships between organizational size and market focus and commitment to the green movement and impacts of organizational culture: A comparative study of Jamaica and the United States. *International Journal of Business and Social Science, 2*(22), 19–34.
Zhang, D., Lu, S., Morse, S., & Liu, L. (2021). The impact of COVID-19 on business perspectives of sustainable development and corporate social responsibility in China. *Environment, Development and Sustainability*, 1–24.

Zhang, X., Chen, X., Zhang, Z., Roy, A., & Shen, Y. (2020). Strategies to trace back the origin of COVID-19. *Journal of Infection, 80*(6), 39–40.

Zouaghi, F., Sanchez, M., & Martinez, M. (2018). Did the global financial crisis impact firms' innovation performance? The role of internal and external capabilities in high- and low-tech industries. *Technological Forecasting and Social Change, 132*, 92–104.

CHAPTER

7

Sustainability: leadership and reporting as its pillars

Diana Fernandes and Carolina Feliciana Machado

School of Economics and Management, University of Minho, Braga, Portugal

7.1 Introduction

The evolution of the concept of sustainable development dates back to the upsurge of environmental consciousness in the 1960s and 1970s (Dovers & Handmer, 1993; Wilbanks, 1994). It has been articulated around three core vectors to present a holistic conceptual approach that entails simultaneously (1) a conservation vision, (2) a community vision, and (3) an economic theory (Hardy, Beeton, & Pearson, 2002; Peterlin, Pearse, & Dimovski, 2015).

Hence in reviewing the literature to provide background, we can observe that the literature on organizational sustainability was pioneered by Benn and Dunphy (2013), and rooted in such an inspiration, it has been advanced by Collins and Porras (2000), Collins (2001), Drucker (2001), Royal, Daneshgar, and O'Donnell (2003), Senge, Smith, Kruschwitz, Laur, and Schley (2008), Mohrman and Worley (2010), Smith (2012), Thomas and Lamm (2012), Florea, Cheung, and Herndon (2013), Wales (2013), and Zawawi and Abd Wahab (2019). Even though these are the most cited authors and therefore provide the most groundbreaking advances to such a conceptual debate, other authors can also be mentioned.

Deriving from such a renewed global motto in the current globalized business scenario, it is increasingly recognized that organizations entail a wider set of impacts in the community in which they are located, expand, and develop. Today they assume a broader and more complex

Innovation and Sustainable Manufacturing
DOI: https://doi.org/10.1016/B978-0-12-819513-0.00005-6

set of responsibilities toward the overall community and environment. Indeed, organizational responsibility has emerged as a concept entailing the "context-specific organizational actions and policies," considering the myriad of organizational stakeholders, the respective expectations, and also the triple bottom line, which comprises the deep and complex connections between the economic, social, and environmental spheres of organizational performance (Arguinis, 2011).

In such a scenario the responsibilities and duties of business as an active part of society are gaining attention, and those impacts extend beyond the mere economic interests of a reduced elite, thus providing momentum to alternative management, economic, and leadership approaches (Freeman, 1984; Hörisch, Freeman, & Schaltegger, 2014; Ricart, Rodríguez, & Sanchez, 2005; Westerman, Nafees, & Westerman, 2021 and, 2010). In the current global economic order, organizations have a myriad of obligations toward a wide scope of stakeholders, which can potentially have conflicting needs, goals, aspirations, and expectations, that nonetheless need to be accessed and considered with the goal of a win-win outcome. Hence organizations today are impelled to act, develop, and expand toward serving the greater good of the overall society rather than merely satisfying the demands of their shareholders (Fernandez, Kullu, & Shankar, 2020; Galpin & Whittington, 2012; Mirvis & Googins, 2006; Peterlin et al., 2015).

Nonetheless, in reviewing the literature, it becomes clear that a fundamental theory of sustainability still has to emerge. Despite the vast interest in the topic of sustainability in contemporary academic and managerial debates, the conceptual rigor is still a milestone to achieve, given that often overlapping and confusing definitions, arguments, and assumptions are presented in the theoretical construct of sustainability (Cabezas & Faith, 2002). Several frameworks have been generated, applied, and developed to identify the vital features of sustainable organizations and how they are led. In this regard, the literature points out that by designing and making use of a cognitive mapping approach, a multidimensional space of sustainability can be deduced to better implement policies, practices, procedures, and tools in the organizational context, allowing corporations to achieve sustainable value chains (Kiewiet & Vos, 2007). Thus according to the literature, financial sustainability, environmental sustainability, and human sustainability are the three vectors that need to be balanced, as they exist in a complex symbiosis to construct a sustainable organization that strives to improve the overall community in which it is located, in all its actions' scope. Hence a quantitative and qualitative analysis, which is clear, for example, in studying the work of Griffiths, Dunphy, and Benn (2003), is fruitful and interesting, since it provides ground for the adoption of the principles of organizational sustainability in every aspect of organizational life.

Sustainability involves a long-term journey in terms of corporate citizenship or, in other words, to a direction in terms of organizational management that requires a sustainable leadership approach and also a rigorous reporting analysis conducted according to the sustainability motto. This will lead to a responsible decision-making process, enabling a full, deep, concrete, and rigorous understanding of the sustainability principles and commitments on the organizational level and beyond via spillover effects.

The current sustainability motto involves consensual demands for green management in the current world. These are increasingly frequent, clear, loud, and urgent, being addressed to a myriad of actors (e.g., politicians, academics, scholars, and managers) and springing from an assortment of sources. They entail laws, national and international guidelines, treaties and regulations, societal mandates, nongovernmental organizations, multilateral agencies, and so on, in which the civil society has been assuming an increasing role by giving intense scrutiny to the business, financial, social, and political worlds (Alfred & Adam, 2009). Based on such a scenario, the truth is that organizations are being impelled to measure their environmental performance, efforts that result from a variety of drivers, including compliance with national and international laws and normative guidelines, ensuring better data for decision-making, supplying accurate and wide-ranging information to regulators and third-party partners, ensuring win-win outcomes to all the stakeholders, and demonstrating progress toward targets, following the arguments of Bennett and James (2017).

Hence organizations are impelled to design, develop, implement, and improve sustainable organizational policies, practices, procedures, and tools so that they achieve high levels of sustainable economic, financial, social, and environmental performance, doing business in a way that respects the recovery potential of the environment and the enhancement of the quality of life of society for both the present and the future generations. In other words, organizations will be improving the overall well-being of the community in which they locate.

This chapter addresses the topic of organizational sustainability, providing insight into the advancements in the literature around green management. It defends a need to rethink academic and managerial debates around such topics, to assume a more ambitious level of awareness in face of the current global dynamics and their consequences for society and the environment. It originates as an effort to provide insight into the designing and application of a model that is capable of stimulating societal development and global environmental conservation by the implementation of organizational policies, practices, procedures, and tools according to a green management perspective. Such discussions require a refined conception of economics that is rooted in the environment and its resources as a core asset in which the relationship between human beings and the natural environment as one of complementarity, not of conflict or even of substitution. This is the core assumption that the chapter articulates.

The argument is structured around four main points that, despite being different from each other and therefore possible to be dissected and analyzed separately, are assumed to be intrinsically connected in their theoretical construction and managerial implications. Therefore the argument will develop as a continuous and logical cadence of ideas so that all sections communicate with each other to show the relevance of sustainable leadership and sustainable reporting as concrete managerial practices in the organizational domain, hence functioning as pillars to effectively and efficiently imprint sustainability as an organizational core value.

The first section of this chapter describes the background of the ecological concern, exposing it as a means to better understand the increasing calls for green management in organizational structures. In this section the historical background for green management will be presented, aiming at defining such a concept as the theoretical basis for the subsequent argument of the chapter. The second section addresses the twofold need and urgency for building credible commitments in regard to the application of green values in organizational management policies, practices, procedures, and tools so as to address and manage the environmental trust gap that may exist and the corresponding consequences of the greenwashing discourse, that is, of presenting ecological arguments to force an ecologically responsible image. In this section, the need to embrace integrated thinking as a means to embed environmental conservation and societal well-being in a business's core concerns will be advocated. The third section discusses a way to ascertain the concept of corporate social responsibility (CSR) as a means to operationalize organizational environmental management and imprint sustainable business value chains. The fourth section corresponds to the corollary of the chapter, defending the urgent need for a multidisciplinary approach to reach insightful and richer debates marked by dynamics of constant analysis and integrated thinking. This section explains that leadership and reporting assume as the pillars of sustainable organizations. Conclusions and suggestions of future research follow.

In methodological terms, the chapter consists of theoretical work, adopting a descriptive exposition comparing and interconnecting different inputs and concepts from various research fields via deductive analysis. The chapter adds to the existing research on the topic by presenting a literature review on organizational sustainability, green management, sustainable development, and CSR, exposing the idiosyncrasies of such concepts but also their connections, which the chapter highlights to point out the relevance of leadership and reporting in organizational contexts as core managerial practices to achieve sustainable business.

Required for such discussions is a refined conception of economics that is rooted in the environment and its resources as a core asset in which the relationship between human beings and the natural environment is one of complementarity, not of conflict or even of substitution. It acknowledges that the concepts of ecological economics, environmental management, green management, sustainable development, and even a circular economy often present conceptual overlap; it also views the concepts as evolving from one other. Thus an evolutive and integrative thinking path can be drawn, showing that all of the above concepts are intrinsically linked, given that they all focus and progress toward the long-term viability of the economy and consider all economic agents to be social agents.

7.2 The ecological concern and the call for green management

In the current global and interdependent world the demands for green management are increasingly being addressed to a myriad of actors, springing from an assortment of sources. Therefore the possible decline in reputation if managers do not display an active and tangible commitment to green management (Sexton, Marcus, Easter, & Burkhardt, 1999) has been assumed as a widespread, consensual, and clear organizational concern in the current era (Alfred & Adam, 2009). Such calls for green management, according to Loknath and Azeem (2017), result from the general awareness of the far-reaching repercussions of human industrial activities on the social and environmental equilibrium, not only threatening the potential of survival, quality of life, and well-being of the current generation but also endangering those of future generations.

Business has always depended on the natural world, and business has always had an impact on the natural world. The critical issue in this symbiosis stems from the fact that the expansion of such a deep and complex relationship has been mistreated and even neglected by economic theory for too long. which has proven to be wrong, because some natural resources indeed assumed as nonrenewable, which threatens the industries' potential of survival, due to resources extinction and difficulties on finding appropriate and efficient substitutes (Loknath & Azeem, 2017).

Furthering this idea, a literature review makes it possible to verify that several studies have considered the concept of ecological sustainability as an updated and fruitful framework to study management policies and practices in terms of both operational and strategic contexts. Thus the concept of environmental management has triumphed as an approach to ascertain the organization's commitment to the social and environmental cause based on the full, deep, and conscious integration of environmental and social concerns into the organization's daily operations, imprinting such topics on the organization's long-term planning and on its quality management systems. As a result of such an approach, transform the organization can be transformed so that it indeed acts in the overall community as a true and complete green system. However, such achievement is possible only if the organization accurately assesses its environmental impacts, sets targets and metrics to reduce these impacts, and plans the means, processes, procedures, and tools by which those targets can be fulfilled.

In connection with the early contributions of Bennett and James (2017), Stainer and Stainer (1997), and Weissman and Sekutowski (1991), we note that Alfred and Adam (2009) postulated that in the current era it is incontestable that a refined enthusiasm exists around the theoretical postulates and managerial models of green management. As a corollary, it can be argued that the vital issue is to understand the rationale by which these expectations to have become so widely accepted in academia, political debates, the management field, business organizations, civil society, and so on (Alfred & Adam, 2009; Babiak & Trendafilova, 2011; Bansal, 2003; Coglianese & Nash, 2010; Delmas & Toffel, 2004; Galeazzo, Furlan, & Vinelli, 2014; Hass, 1996; Lamond, 2009; Sharma, 2000; Sharma, 2020; Vogel, 2005).

Bennett and James (2017) produced an analytical model that provides eight sequential steps for effective and efficient environmental performance appraisal: (1) select appropriate measures, (2) set targets, (3) identify potential measures, (4) implement measures, (5) define the environmental context and objectives, (6) monitor and communicate results, (7) act on results, and (8) review. They present the organizational commitment toward the social and environmental cause as moving along a spectrum that develops in the form of a continuous loop via a gradual process of increasing engagement levels. Correspondingly, and as a review and further development of such a proposal, Stainer and Stainer (1997) suggest that the idea of creating a performance cycle process could be understood as, at its very core, corresponding to an ongoing dialog and a pivotal point across the organizational structure via a transversal dynamic between all stakeholders.

There is a consensus in the current literature that as a result of globalization, the business sector has recognized and, most important, legitimized the prerogatives and the claims of green management, conceiving them as not just a hot topic and a social trend, not just a defensive instrument for retaining and strengthening organizational legitimacy and increasing organizational credibility, and not only to fortify the organizational compliance to national and international regulations and its right to operate as a business entity, but indeed to act as the nucleus of an organization's ongoing mission and reason for existence, growth, and development.

According to this logic, the premises of Alfred and Adam (2009) are relevant, as the authors advocate that sustainable development consists of the core conceptual umbrella to understand the theoretical construct of green management as it deeply connects to such a concept because, at its core, it pursues a harmony among three vectors (i.e., environmental, social, and economic). Nonetheless, it is vital to point out that such harmony is difficult to achieve, given that inherent to organizational actions are the often conflicting interests of a myriad of stakeholders. Loknath and Azeem (2017) defend the articulation of the three core vectors in terms of the theoretical construct of sustainability, postulating, similarly to Alfred and Adam (2009), that the three symbiotic extents relating to the environment, economics, and society act as a triple bottom line, advocating that the overall business industry practices need to urgently and widely embrace the notion that sustainability derives from focusing on the interdependent dynamics proposed by the holistic perspective that this triple bottom line defends (Slaper & Hall, 2011). The term's origin dates back to the mid-1990s, and it corresponds to an evolving approach to understanding the connection between business and the natural, social, and economic environments; as each is reconfigured, the others also change. The essence of such a concept addresses a holistic, open, wide-ranging, updated, and systematic approach in terms of the management of the complete arrangement of an organization's duties, tasks, actions, and responsibilities. By the expansion of such an assumption, the concept is used, in a broader significance and application, to assess, capture, expose, and evaluate the complete set of values, goals, matters, tools, and processes that organizations must address and manage so that they can efficiently and effectively maximize the potential positive consequences of their activities in the several domains in which they operate. Thus they effectively generate added economic, social, and environmental value, strengthening, sustaining, and greening their value chain. Subsequently, this approach suggests that organizations focus on adding value simultaneously on those three vectors, which implies that they should also be publicly accountable along the same three vectors, abandoning the conventional financial myopia analysis. As a corollary, we postulate that the triple bottom line approach tries to understand how organizations assess and manage all the three responsibilities. By doing it as a first effort, it develops in its analytical potential to account for these interrelated scopes of organizational activity. Hence and as a vivid advantage, this approach can provide insight into a more balanced interpretation of the overall organizational performance.

Having already provided the historical and conceptual background to understand the concept of green management, we now provide a clear definition for it to make it understandable and operational (Starik &

Marcus, 2000). According to Haden, Oyler, and Humphreys (2009), green management consists of a refined organization-wide process that aims to apply ongoing innovation to achieve sustainability, waste reduction, environmental conservation, social responsibility, societal progress, regional development, enhanced quality of life, and a competitive advantage resulting from the continuous learning and development process. Thus the authors' theoretical suggestion for the green management concept postulates that the inherent goals can be accomplished by embracing environmental goals and strategies that are fully integrated with the other goals of the organization. These ideas reinforce the core notion of organizational commitment, and indeed a review of the literature shows that researchers have identified three subcomponents: (1) effective commitment, (2) continuing commitment, and (3) normative commitment. In this sense, for example, Cairncross (2014) establishes equivalents between striving for overall organizational quality, organizational prerogatives, and efforts in terms of environmental management.

This more comprehensive and updated definition of green management is inspired by the earliest contributions set forward by Taylor (1992), who claimed that green management would be "the next competitive weapon" in the global business organizational environment, based on the premise that the environment was indeed a focal point to change the self-image of organizations and to create new competitive advantages. On the same thinking path, we can postulate that, for example, Makower's (1994) insights are inspired by Taylor's (1992), presenting a correspondence to the ideas set forward by the above-mentioned authors, as Makower (1994) simplifies the spirit of being "green," from the organizational part, as concerning two different but interconnected elementary goals reducing waste and maximizing resource efficiency, advocating that updates are fundamentally required on the operational part of the organization and introducing four key concepts as guidelines to review process design and implementation so that the organization can improve its overall performance and achieve positive results both in the economic, social, and environmental domains in which it operates: (1) empowerment, (2) education, (3) efficiency, and (4) excellence.

Inspired by the work of Makower (1994) and furthering it, a literature review on this topic shows that the most frequently mentioned positive consequences of the application of green management theoretical principles and managerial models to organizations may be congregated into ten broad classes: green operations, leveraging profitability, improved operational safety, enhanced resources utilization efficiency, lean management, effective operations, competitive product/service, market expansion, improvement in company image, and improvements in management and leadership (Hart & Ahuja, 1996; Weissman & Sekutowski, 1991).

Taylor's (1992) proposal refers to a management concept that requires the pledge of all organizational stakeholders, so it entails a comprehensive and holistic concept that involves conceiving of the organization and its business value chain in the long term as an active and perfectly integrated part of the community. Subsequently, this concept aims at managing the entire system, which implies addressing no conceptual distinction between structure, process, production, operations, and administration.

Following this line, Taylor's (1992) conceptualization of green management entails observing the organization in its entirety rather than as a mere collection of smaller, diversified, and separate entities, managing the organizational domain toward the long term. As a synthesis, it is possible to observe a correspondence between the perspectives of both authors, as they all advocate that the organization exists as a part of the overall social and environmental community.

In respect to the above-mentioned perspective and systematizing some insights that we provided in the first paragraphs of this section in providing background to understand this concept, Taylor (1992) summarizes the reasons for adopting green management approaches, stressing the relevance of (1) marketing opportunities; (2) cost savings; (3) enhanced the competitive position; (4) improved employee motivation; (5) generating new opportunities in processes, goods, services, tools, procedures, and structures; and (6) the enhancement of a good corporate image.

Not disregarding the insights so far exposed, the truth is that it is still difficult to provide a clear, consistent, comprehensive, and consensual definition of the term *green management*. The literature presents vague, confused, overlapping, and even incomplete definitions. One of the primary reasons derives from the fact that it is a relatively recent term (e.g., Banerjee, 2001a; 2001b; Freeman and Kunreuther, 2002; Haden et al., 2009; Hass, 1996; Siegel, 2009; Hörisch et al., 2014). According to Haden et al. (2009), in addition to the above-reported difficulty, another obstacle exists, given that many other terms are frequently used as synonyms in referring to green management when at their very core they refer to different, albeit interconnected, concepts. Here, we mention terms such as environmental management, CSR, corporate environmentalism, sustainable management, and circular economics (Ghobadian, Viney, Liu, & James, 1998; Jennings and Lumpkin, 1992; Rothenberg, Maxwell, & Marcus, 1992).

As an oriented managerial operationalization of the theoretical postulates of green management, we shed light on the analytical model suggested by Hosseini. We suggest that it is possible to constitute a connection, as a source of inspiration, between such a model and the insights previously highlighted by McWilliams, Siegel, and Wright (2006) and by McWilliams and Siegel (2011), given that the authors emphasize, as issues to consider in strategic environmental social responsibility, the vectors of: (1) information asymmetry, (2) product differentiation, and (3) advertising. Thus Garzella and Fiorentino (2014) and Siegel (2009) end up by summarizing that recent research indicates that an organization's decision to engage in green management, from here deriving the organization's pattern in terms of its engagement level materialized into CSR, corresponds to a true strategic choice, which can and should be constantly revised via an ongoing process, keeping in mind the different configurations of the overall community in which the organization acts, grows, and develops. Siegel (2009) advocates that both individual and organizational factors are pivotal in understanding the antecedents and consequences of these decisions and its materialization in terms of actions.

Given the above background context, the analytical model proposed by Hosseini in terms of implementation of green management in the organization's daily activities is articulated into four vital vectors, which despite differences, interconnect as they form an open and complex system. Those four vectors consist of (1) inputs, (2) processing and operations, (3) outputs, and (4) other affected variables.

The system's inputs entail raw material, energy consumption, labor, ergonomic machinery and equipment, and tools as well as information, data, and knowledge. In regard to the processing and operations, the model provides further insights involving the relevance of design, material handling, mismatched facilities, safety performances, proper techniques, and coordinated work as well as good utilization of facilities, presenting such an enumeration of details around the organization's processing as core features to make its overall activities "green." The model then highlights, as outputs, the improvement of the planned services, the need to enhance high service quality, the constant focus on information and knowledge creation and sharing, and the assurance of security and healthy for citizens and customers as well as for leaders, managers, and employees. Concerning the other affected variables, the model explains that those aspects involve the core notion of green culture, which should be part of the organization's daily actions, as can be observed in the daily use of sustainable and socially and environmentally friendly materials, techniques, and tools; in the organization's constant concern with its external and internal impacts; on its proactive compliance with national and international regulations; and on the organization's ongoing focus on the sustainability of its value chain and benefits systems.

Such a model supports imprinting the green management postulates in the daily operations of an organization to establish strategies and managerial approaches framed upon a goal-oriented model that as a prerequisite entails the connection of three basic dimensions: (1) environmental considerations (specifically, paying attention to national and international environmental standards, to governmental policies, to the actions and guidelines of nongovernmental agents, etc.); (2) productivity, which in other others can also signify a concern for reducing cost (specifically, paying attention to the allocation and use of raw materials, energy, information, and technology); and (3) customer's needs, goals, and expectations (specifically, paying attention to security and health safeguards in products, services, and production processes; to product and service design; as well as to the communication and interaction channels and flows).

7.3 Building credible commitments: addressing and managing the environmental trust gap

According to Loknath and Azeem (2017), the meaning of the term *green* can be applied to a myriad of fields, as it entails considerable scope articulating a kaleidoscope of vectors. Such a management approach aims to be a mindset that suffuses the entire organizational structure, policies, practices, tools, and procedures and not a mere add-on or unique feature. To fully design and implement such a management approach in any organization requires a full and deep strategic transformation of the organization's management and leadership mindset, policies, practices, tools, and procedures, based on the assumption that the ongoing, flexible, open, and dynamic process of greening the economy (the production processes, the business value chains, and the organizational structures) not only pursues the production of goods and services via more efficient processes in regard to resource allocation and waste spillovers, but also aims to produce goods and services that support a greener outcome.

Green management refers to a holistic management approach that conceives of the business and natural environments as fully interconnected, emphasizing the need to have elevated performance standards in terms of environmental management, focusing attention on greening the inputs, outputs, production, distribution and storage processes.

The climax of such an argument postulates that in the current times, it is more than consensually accepted that green management consists of a management approach that will be economically and financially profitable (Darnall, Jolley, & Handfield, 2008; Porter & Van der Linde, 1995; Raufflet, Bres, & Filion, 2014; Sharma & Vredenburg, 1998; Sharma, 2000; Sharma, 2020). Considering such management standards concerned with environmental preservation, resource sustainability, and human well-being, this approach is assumed to be a driving force for corporate entrepreneurship and innovation (Cohen & Winn, 2007; Dean & McMullen, 2007; Haldar, 2019; Koe & Majid, 2014; Matthews & Senyard, 2010; Muñoz, Janssen, Nicolopoulou, & Hockerts, 2018; Rajasekaran, 2013; Sharma & Ruud, 2003), undermining the conventional premises of corporate management that claimed that green management imposed additional costs, thus slowing productivity growth the subsequently hindering efficiency and competitiveness.

We consider that by dissecting this topic, it is possible to ascertain the trust gap issue. First, we focus on the concept of trust and its application in the corporate management domain. In the organizational domain as in any other, trust should not be understood as an ethereal and predefined concept; it cannot be defined per se, it can be defined comprehensively only by the so-established relationships between a given organization and all its stakeholders, all of them with their idiosyncrasies. Thus, even though the concept of trust can be defined as abstract and subjective, which hinders its analytical potential, it can indeed be understandable and actionable.

We need to interconnect the concepts of trust, credibility, intent, and competence, as each of them, despite differences, often overlap, rooted in the premise that building trust, especially across a wide range of stakeholders and imbedded on the corporate scenario, requires a proactive, holistic, thoughtful, and structured approach. Indeed, credible climate commitments are trustworthy and therefore can be designed, implemented, and achieved with an expectancy of success.

Organizations build and maintain trust with all their stakeholders through their actions, and the key premise is that this trust is manifested via an aware correspondence between the organization's competence and intent, which results in organizational credibility and legitimacy. Therefore organizations need to fully interconnect competence (their ability to execute) with their intent (the reasons or motivations behind the organizational actions), designing and implementing a true foundational basis for trust, which then mirrors fairness, transparency, honesty, learning, constant improvement, enhanced organizational performance, and overall environmental and societal impact.

We can go further by postulating that this trust gap concern emerges from the very essence of green management in both its theoretical construct and managerial approach, given that it fully entails the balance between inputs, production/distribution/storage processes, and outcomes, with equal relevance in terms of environmental and performance standards. This develops into the argument of Alfred and Adam (2009), which postulates that, irrespective of which vector the analytical focus sheds light on, an efficient and effective green management approach in any organization conceives the natural environment and the business scenario as connected parts of a whole complex system. They cannot be dissociated, so actions in each of them influence the others. Such a twofold theoretical and managerial understanding embodies a steady essence in the definition of green businesses, postulating that they will be able to be efficiently and effectively observed only in an organization that is fully, actively, and consistently committed to the principles of environmental sustainability in the broader kaleidoscope of its structures, policies, procedures, processes, tools, assets, and operations, always bearing in mind the constant concern to minimize the negative environmental impact of the organization's overall activities, which are managed by high-level sustainable standards. In this perception the organizational efforts strive to achieve environmentally friendly business value chains, and therefore sustainable competitive advantage, which suggests that the management concerns are directed toward the long term, conceiving of the organization as a true part of an overall system formed by a joint triad of vectors, namely, economic, social, and environmental, understanding that organizations have an important role to play in the survival potential, quality of life, and well-being of both current and future generations.

Loknath and Azeem (2017) agree with the previously mentioned authors and address the relevance of this trust gap as part of the essence of the green management approach by advocating that green businesses consist of organizational structures and business value chains that adopt green management structures, principles, policies, processes, tools, and practices, designing and executing them to try to satisfy the interests of all the involved stakeholders' via concerted negotiation, building common points and aiming at the common good as well, thus improving the quality of life for their customers, managers, vendors, employees, the communities in which they operate and develop, and the environment in which they are located.

Going further into the trust gap problem, we hereby state that its fundamental essence derives from the assumption, as demonstrated by Alfred and Adam (2009), that businesses will not necessarily introduce green management practices because of the normative obligation that may be inherent in them. In this view, environmental management concerns are seen almost exclusively as a threat and/or as a legal obligation laid on the organizations by the multiplication of environmental legislative instruments and agencies. However, in a contrasting view, organizational structures and business value chains are likely to aim at designing and implementing green management approaches because, in their organizational culture, vision, and goals, green management coincides with their economic interest in satisfying all the key stakeholders and thriving in the current global business environment as profitable companies, with competitive advantages that are sustainable in the long run.

In other words, the trust gap emerges from the twofold need and goal to conceive the outcomes of the organizational development as a win-win situation for all the involved stakeholders. Indeed, in such a management approach, win-win solutions are designed and materialize with the aim of achieving sustainable standards, societal well-being, regional development, business purposes, corporate performance levels, and organizational growth goals, attributing to all of them an equitable relative relevance in the overall business system (Hart, 1995). Siegel (2009) climaxes this argumentation by asserting that organizations, foremost focusing on the ones which comprise corporate business structures, should adopt green management structures, processes, procedures, tools, and practices only if those effectively and efficiently complement the organization's business- and corporate-level predefined strategies, this also to say, if those as well ultimately leverage the organizational business profitability, enhance its business value chain, and shareholder wealth.

Addressing the trust gap between the theoretical premises and the managerial materialization of the green management approach entails focusing on building credible climate commitments, which can act as a road map to earning all the organization's stakeholders' trust in the design and implementation of such an approach so that the inherent climate pledges are addressed and fulfilled by engaging in ambitious commitments that, at their essence, prove to be authentic to the organization's purpose, values, and mission with regard to the overall natural environment and societal milieu and hence become deeply grounded in the organization's business strategy over time.

It can be posited that the origin and goal of this trust gap and the need to address them results from the grounding assumption that the stakeholders' expectations with regard to climate are on the rise, which has resulted in businesses facing critical legitimacy and credibility challenges, especially in the current era marked by the widespread impacts of globalization. Addressing, managing, and overcoming such challenges to corporate legitimacy, credibility, competence, and trust entails considering all the stakeholders' expectations collectively, acknowledging that they consist of a myriad of interests, needs, and goals. While each of these organizational agents has its idiosyncrasies, having a general understanding of how these agents conceive and understand climate and societal commitments can indeed help companies to design, manage, implement, and communicate climate and social policies and actions effectively. Therefore all the stakeholders' different and potentially colliding pressures can best be managed and overcome via a win-win strategy that promotes the capacity of the organization to overcome organizational challenges that are diverse, often unexpected, dynamic, and evolving.

We should also point out that addressing the trust gap by incorporating environmental and social concerns into corporate management requires ensuring that such environmental and social commitments are firm and credible, rooted in a shared awareness among all the organization stakeholders and a common goal. This becomes arguable when we consider that greenwashing (either being real or perceived) can lead to multiple negative consequences, which can indeed spread from the corporate domain to spill over to the overall society, thus eroding the public image of the organization, affecting its legitimacy and credibility (and therefore trust in the organization itself and its actions).

In such a scenario, stakeholders may question the organization's capabilities, competence, and/or intent to reduce its climate impact and to develop a positive impact on overall societal well-being. This situation can have a negative impact on the credibility of the organization, which can vary along with a spectrum and develop into a myriad of spillover effects, such as causing actual or potential partners and consumers to reconsider acquiring the company's products and services, which may entail searching for substitutes if the organization does not fully meet the desired performance standards in regard to environmental management and societal well-being. Among other consequences, this may lead to a struggle, on the organization's side, to attract and retain top talent and may in such a scenario cause higher levels of turnover as the employees do not feel that the organization lives up to their own environmental values (Wesselink, Blok, & Ringersma, 2017). Continuing on such a spectrum of negative consequences, these harmful impacts may increase the company's cost of capital, hence threatening the allocation of capital to its present and future investments and growth plans. This may affect the organization's attractiveness to investors, which in the end can result in a threat to its viability as a business and even its societal license to operate.

Thus emerges the urgency of designing and implementing an integrated thinking model as an operational tool to design and implement an efficient and effective roadmap toward credible and legitimate organizational climate commitments. Very important is the premise that such an approach consists of a process and not a destination. It develops in a temporal spectrum as a continuum of policies, practices, and tools in which the organization can configure itself, varying with the stakeholders' expectations and the organization's goals. Adopting such a mindset and taking the next step to implement it require internalizing a recurrent process that constantly monitors and evaluates progress so that not only are competence, intent, credibility, legitimacy, and trust among all the stakeholders clarified, communicated, and engendered in the present scenario, but also active steps can be taken to enhance them in the future.

To better understand such an idea, we will focus on the concept of *integrated thinking*, which involves a holistic and cohesive management approach to fully and deeply incorporate sustainability as the business's core, helping organizations to understand the dynamics and challenges of the value creation process, enhancing the sustainability of their business value chains through a new lens that proves efficient and effective in providing equilibrium between short- and long-term outcomes, as it acknowledges the kaleidoscopic range of resources (natural, social, and financial) that are available and upon which organizational growth, development, and survival rely. Therefore conceptualizing and implementing such an approach require a mindset that crystallizes both theoretical considerations and managerial applications that make viable the process of embedding climate considerations and societal concerns into every facet of the organization's structure, policies, practices, tools, and procedures via an approach that links different but connected fields of organizational action, such as strategy, risk management, governance, communication, targets, and metrics.

As concrete and proactive steps to address the trust gap, we support the design and the implementation of the following on the part of the organization as effective and efficient means to integrate a wide-ranging purpose and attention to the environment and society into their actions and, in turn, boost trust with stakeholders:

- ***Establish the foundation***: This means that organizations should first make efforts to understand their current environmental and societal impacts, articulating their direct and indirect actions and agents, and assessing the myriad of forms on which they can exert influence. It is vital for organizations that wish to take a credible environmental and societal approach to sustainability to critically evaluate their existing reporting processes, practices, and tools to proactively identify current and potential stakeholder and regulator expectations. These efforts will later allow them to best assess and manage those pressures.
- ***Commit to action***: All of an organization's agents and structures, using its leaders as role models and mentors, should put into place ambitious climate goals that have the visible and vocal support of all the organizational actors, including (foremost) the organization's board.
- ***Strategy***: As a corollary, we can postulate that all the efforts result in the revision of the organizational strategy, imprinting on it a truly environmental and societal concern, so that we can observe the development of goals with an appropriate, aware, and clear focus, leading to appropriate, aware, and clear targets that truly embody the characteristics of the most robust climate and social efforts, thus enhancing legitimacy and credibility in both the overall business environment and civil society, bridging the potential trust gap. To objectively materialize such efforts, the organizational perspective and goals must be aligned in terms of climate impacts, environmental preservation, and social well-being with the current developments concerning the scientific consensus about what is needed, thus making it possible for the organization to work with third-party actors to generate a consensual, coherent, and vigorous starting base for policies and actions. The strategy should involve all the

organizational structures and actors via a joint, coherent, and equitable dynamic, fostering in all of them the full information, awareness, empowerment, and will to redesign and actively transform every part of the business (operations, supply chains, talent, logistics, research and investigation, product and service innovation, delivery excellence, human resources management, information technology, and beyond) to fully embed and embrace social and environmental sustainability. This will lead to the greening of the value chains, incorporating clean technology and circular processes. Efforts toward such goals may impel the organization to access new products, to redesign processes, or even to take part in new projects, which may result in new investment needs, augmenting the costs the company may face. Therefore organizations should be aware of such needs and face them and development and growth opportunities, which may enhance their competitive advantage.

- *Governance*: Organizations need to imperatively design and implement clear ownership and accountability for their environmental and social commitments. Therefore all organizational actors need to be appropriately informed, educated, empowered, and willing to assess, address, and champion those efforts, acting as a catalyst for the greening of the overall organizational footprint. To drive the necessary changes, it is first necessary to assess whether all the organizational actors would be able to access the needed tools and opportunities to develop their environmental and social awareness, to dedicate to such actions and initiatives, and to increase their knowledge of such topics. Therefore forging credible and robust organizational climate commitments and overcoming the potential inherent trust gap, would require establishment of organizational concrete metrics at the business unit, value chain, team, and individual levels and geographical/regional scope where appropriate to enable a condensed action and effort aimed at incentivizing the climate and social action by fully and deeply embedding it in performance management and remuneration policies as a true component of the organization's values and mission.
- *Metrics and targets*: The expansion of the previously mentioned efforts also requires that reporting methods, processes, policies, and tools be deeply and critically reviewed and then aligned with leading third-party standards and frameworks to enhance their credibility and legitimacy in civil society and their attractiveness as sustainable business standards and goals. This means the designing and adoption of a long-term perspective, conceiving the strategies and actions plans derived via a holistic perspective, given that only such an approach truly corresponds to the sustainability concept.

Subsequently, this idea postulates the need for the organization to integrate sustainability, social well-being, and climate conservational measures into its internal audit plan as well as evaluating the type, action path, strategy, process, and level of external assurance essential to meet all its stakeholders' expectations. If we go further in such an argument, it can also be said that to enhance such a process and promote its outcomes, it would be very useful for the organization undertake benchmarking against peers on select industry-comparable metrics, which could enhance the credibility and legitimacy of their environmental and societal commitments, thus best managing and overcoming the trust gap.

- ***Risk management***: The need to address specific risk management approaches so that the inherent organizational environmental trust gap can be best managed stems from the assumption that the contemporary landscape of global climate commitments is altering rapidly and deeply, reconfiguring into new standards, processes, goals, timelines, and tools as society, organizations, and institutions accelerate and loudly express their response to the urgency of the climate and social crises. This to say, the relevance of designing risk management approaches to build credible and robust environmental and social commitments requires understanding on the organization's part of both the internal and external environments (and also the one formed by the connection of the two). Such risk management approaches require a comprehensive perspective when they are designed and implemented, which means that they can only be expressed in concrete terms when they are framed into a holistic perception, given that building trust, legitimacy, and credibility with key stakeholders requires credible plans, needs, goals, performance standards, and milestones as well as assessing, measuring, evaluating, and constantly, openly, and freely communicating them and their progress to all the organizational stakeholders (including the overall public and their business colleagues), so that they do not face reputational risks and can actively be part of the overall business dynamics.

Now is the right time to accelerate such a twofold process and progress. A synthesis of the above exposition makes it possible to ascertain that there is no overall standardized rule to define and implement a sufficient corporate environmental and social commitment. Nonetheless, such commitments, if indeed aimed at bridging the possible trust gap as explained earlier in the chapter, would need to entail a symbiosis of key attributes, namely, a comprehensive scope, science-based targets, and a rigorous assurance of their dynamics and relevance. Only in this way will credible, robust, clear, ambitious, rigorous, and authentic climate commitments become possible. Only with such a mindset can organizations, businesses, and the overall society embrace integrated thinking and embed environmental conservation and societal well-being into their core concerns.

- ***Comprehensive scope***: The organizational efforts to build credible and reliable environmental and societal commitments should be designed and implemented to encompass as much of the organization's climate and social impact as possible, entailing all the organization's structure, policies, practices, tools, agents, geographical areas of activity, business chains, industry and business sectors, operations, and operational sectors, subsequently covering all its direct and indirect actions and impacts. Thus those efforts shall emerge by a holistic approach so that the inherent trust gap can be avoided at the very beginning.
- ***Science-based targets***: The organization's goals and respective performance standards should be designed, operationalized, and assessed in a common understanding, aligning all the stakeholders' interests, needs, and expectations in a win-win approach embedded in the current science-based advances to further legitimize such ambitions and the respective strategies and operational models. As a corollary, the targets should be focused on the most important outcomes for addressing climate change and societal well-being.
- ***Rigorous assurance***: This vector at first implies that climate commitments should be conceived, designed, implemented, assessed, and improved as being clear, fair, honest, and transparent, which results in the assumption that organizations need to both report and disclose accurate, constant, updated, and detailed data about their current and past footprint, inherently also committing to report on progress, revision, and updates openly and regularly. The idea culminates in the argument that the environmental and societal data that are generated by the impact of the organizational actions and the respective performance shall be capable of being captured, measured, evaluated, and validated not only by the organization itself but also by third parties that follow leading standards. The third parties may be other organizations in the same industry, business partners, competitors, adjacent businesses, scientists, politicians, scholars, civil society, nongovernmental agents, and so on.

We do not wish to present a negative perspective on the topic. Nevertheless, while commitments that reflect these three attributes might be considered a leading practice in the design and implementation of credible, ambitious, and reliable environmental and societal commitments on the organizations' part, they represent the bare minimum of what is required if we are fully aware and indeed willing to engage in truly limiting catastrophic climate change and the surrounding social crisis.

7.4 The concept of corporate social responsibility as a means to operationalize organizational environmental management and imprint sustainable business value chains

According to Loknath and Azeem (2017), various researchers who have been dedicated to the domain of green management have traditionally categorized the concept of CSR from a kaleidoscope of assumptions, which can be better understood in light of the classification of organizations based on their strategic options. Therefore we can, as a starting point, discuss the difficulty of a consensual initial definition for this core concept of CSR in terms of the organizational environmental and social footprint. This stems from the fact that, as has been pointed out in the literature on this topic, environmental management and corporate sustainability have traditionally been investigated as concepts applied in close conjunction with (or even as a substitute for) green management. The literature clearly points out the theoretical, conceptual, and managerial overlap, that is, fusion and confusion.

Nonetheless, in the literature it is seen that both concepts seem to entail a wider and more significant scope It can be arguable that they more accurately and fully embrace the idea of green management when compared to the fragmented and initial description of corporate environmentalism offered by Costello (2008).

Indeed, the nucleus of the environmental management theoretical construct focuses on management systems that have been designed, implemented, and developed to achieve continuous improvement to act as an effective approach for proactively dealing with environmental issues (Kautto, 2006; Kivimaa and Kautto, 2010). Hence they also entail an economic conceptual vector as it crosses with the competitiveness and sustainability of the business value chain that the organization presents. Therefore it affects organizational performance and can also be defined in terms of economic profit (Dresner, 2012; Florida, Atlas, & Cline, 2001; Goodland & Daly, 1996; Hervani, Helms, & Sarkis, 2005; King & Lenox, 2001; Lefebvre, Lefebvre, & Talbot, 2003; Puopolo, Teti, & Milani, 2015).

From this comprehensive viewpoint, the concept of corporate sustainability, which requires continuous improvements in the organizational structure, strategy, policies, processes, and tools to achieve its challenging objectives, may evolve into the concept of CSR as a core value, goal, and action so that the organization thrives by achieving those aims. Furthering this idea and according to Stead and Stead (2000), transforming sustainability can be seen as an accomplishment instead of a mere goal. This implies that the contemporary economy must urgently, drastically, and fundamentally shift from a focus exclusively on economic growth to involve active steps to design and implement a steady-state economy that allows the survival and good quality of life of the present generations without risking the survival potential and well-being of the future ones (Hawken, 1993). This view results in the conceptual umbrellas of green management and circular economy (Fernandes & Machado, 2020a, 2020b). In other words, such a resource management approach can make it possible that the rates of pollution emissions do not exceed the assimilatory capacity of the earth and that the levels of consumption do not surpass the levels of regeneration so that the rates of nonrenewable resources do not exceed the levels at which sustainable substitutes can be found, developed, and used (Azzone, Noci, Manzini, Welford, & Young, 1996; Crane, McWilliams, Matten, Moon, & Siegel, 2008; Daly, 1991; Foote, Gaffney, & Evans, 2010; Harris & Crane, 2002; Johnson, 1998; Lee & Ball, 2006; Netherwood, 1996; Orlitzky, Siegel, & Waldman, 2011; Taylor, Barker, & Simpson, 2003).

Such an approach requires a mindset that conceives of protection of the environment both as an anthropocentric moral and ethical issue and as a means of leveraging organizational performance and its financial benefits. The concepts of competitive strategies, and competitive advantage, and foremost their sustainability result from such a theoretical and managerial approach.

To better understand the concept of CSR and its operational value to materialize environmental management and green management paradigms in organizations, directing them toward a circular economics, it may be useful to review the literature to find insights concerning different perspectives on corporate responses to social and environmental pressures.

In an earlier contribution, Simpson (1991) categorizes corporate responses to environmental pressures in three main groups: (1) the "Why Me," (2) the "Smart Movers," and (3) the "Enthusiasts." The author explains that the first group consists of companies that have been forced to expand, enhance, and improve their environmental performance, such pressure being felt via a gradual process as a result of a given well-publicized event. Therefore such pressure and the actions that are taken to achieve those goals aim at improving the public image of such an organization; therefore those efforts result from the catalyst event or situation that induced the company to take some action in the environmental and social fields. The second category corresponds to the business agents (organizations and their actors) that have been able to actively and successfully exploit the opportunity created by the arrival of the "green consumer" to the markets by designing and implementing a myriad of efforts to gain, develop, and sustain competitive advantages. The third group may be understood as including fully aware and active environmental and social business agents representing the more ambitious, robust, and coherent CSR approach because such group includes organizations that have moved beyond compliance and do not just develop according to what the law dictates but instead fully embrace the social and environmental concerns in their overall business strategy and organizational structure along with their present and future actions.

Roome (1992) presents five environmental strategies on which organizations can rely to effectively implement environmental management approaches. The author divides such strategies into a spectrum of increasing engagement and commitment toward green management, therefore postulating that the organization's CSR policies and practices can manifest from being noncompliant, compliant, compliant-plus, providing commercial and environmental excellence, and being at the leading edge. Thus Ketola's (2008, 2007, 1997, 1993) argument inspires the above-mentioned contribution, and the author goes further to crystallize such CSR policies and practices into organizational strategies that can be observed along with a spectrum of gradual ambition and commitment toward the social and environmental fulfillment, these being referred to as stable, reactive, anticipatory, entrepreneurial and creative in a strategic posture analysis.

The suggested spectrum presupposes a progressive evolution in terms of the environmental and social strategies of the organization, that is, in the scope of its CSR, because as the organization advances in the several categories of the spectrum, it also advances in commitment and ambition levels toward green management, which implies looking beyond the existing standards to conceive the integration of environmental management topics within the entire management system of the organization via a dynamic entailing all its structure, policies, processes, and stakeholders. Only with such a mindset will the organization be able to fully recognize and capitalize on the opportunities that are created by the environmental revolution and corresponding social progress. This suggests that those organizations will actively strive toward state-of-the-art of environmental management and overall societal development, gaining, sustaining, and leveraging the organizational competitive advantage.

James's (1994) framework is similar to the one postulated by Roome (1992). The author defines four categories that represent the different environmental strategies that organizations can use concerning their engagement and commitment toward social and environmental concerns, developing a thinking path very similar to that of Roome.

Beaumont, Whitaker, and Pedersen (2013) developed a categorization in which corporations are placed at six different levels according to their response to the environment. The first two levels are similar to the categories of Roome (1992) and James (1994) determined by the organizational noncompliance and/or compliance toward social and environmental standards, but it is in those levels that the new model presents its specificity and originality. Indeed, the third level in the categorization of Beaumont et al. (2013) is known as corporate action, postulating that organizations that design and implement environmental strategies and CSR policies under this category are the ones in whose structure the management team begins, in a fully aware effort, to conceive of social and environmental topics as relevant and urgent, perceiving such issues as a twofold business challenge and opportunity, developing efforts to embrace a holistic perspective of the environment. The authors consequently view their model as a progressive path in turning the engagement potential of the organization toward a more ambitious CSR, along with a spectrum in which the commitment level of the organization toward its social and environmental footprint increases. Beaumont et al. (2013) postulate that at the fourth level of the model, objective changes will take place in the organization's structure, strategy, policies, tools, and processes that will crystallize as an organizational response to the social and environmental issues that have been increasingly addressed by the business environment and by the overall global civil society. The fifth level corresponds to supply chain action, in which several social and environmental problems undoubtedly are an integral part of the entire industry's supply chain and so become a critical vector of the organization's structure, operations, growth, and development. At the final level of the model, the authors postulate the business scope action, in which the organization actively expands its activities by making use of social and environmental matters to gain markets, to gain and sustain its competitive advantage, and to get ahead in business.

Steger (1996) provides a useful contribution on this topic, inspired by the proposals previously set by Roome (1992) and James (1994) and reviewing and developing them. The difference in the conceptual model resides in the fact that Steger (1996) perceives the variety of corporate responses to the social and environmental problems as being merely based on social and environmental risks and on market-based opportunities and therefore on a pure game of calculations to capitalize on investment opportunities and the corresponding revenue potential, relying on pure financial logic. In the models of Roome (1992) and James (1994), the factors elucidated by Steger (1996) are relevant but are not exclusive to developing a comprehensive organizational approach toward social and environmental concerns. These authors advocate that equally important in symbiosis with the previously exposed factors are vectors such as normative standards and guidelines, national and international legislation, constraints within and exterior to the firm, and the configuration of the organization's workforce, explicitly the ability of leaders and managers to create awareness and to bring about organizational change to integrate social and environmental issues.

Not disregarding such differences, Steger's (1996) model is in line with the ones postulated by Roome (1992) and James (1994) because it classifies the diverse corporate strategies into four categories in regard to their commitment toward social and environmental efforts: indifference (which represent the organizations that display reduced environmental risk and indeed fewer and merely superficial environmentally based opportunities for growth), offensive (involving organizations that show a substantial potential for exploiting environmentally related market opportunities), defensive (involving the organizations from the business sector that display a higher environmental risk and hence cannot merely ignore environmental issues, because that would threaten of their survival potential in the current global business scenario), and innovative (involving the organizations that have increased levels of environmental risk in their everyday operations but that simultaneously have numerous and various environmentally based opportunities for growth and development in the current global business landscape).

Welford's (2016c, 2016b, 2016a, 1998, 1995, 1994) categorization of small- and medium-sized enterprises into four main groups corresponds to the logic so far explained. Nonetheless, it is slightly different because it applies to a very specific type of companies. In the author's model, the first group of companies is referred to as the "ostriches" in terms of their commitment to the social and environmental footprint, that is, to the engagement potential of their CSR. Explaining the label, the author suggests that those organizations assume that the current concern for the society and the environmental cause consist of a passing phase, a mere topic trend at a given time. They actively defend the impact of their business activity on the environment and the society as negligible at best. Such organizations also assume that their competitors have the same mindset and hence make no effort to address the social and environmental problems, as these will be no topic for their agendas. The second category of the model categorizes organizations as "laggards" in terms of their CSR approach. In such a category are companies that are aware of the social and environmental challenges but assume that they are unable to combat and overcome those problems, mainly because of cost constraints on their financial balance, lack of a trained workforce, an unprepared management team, lack of knowledge, rudimentary technologies, and so on. The third group of organizations consists of the "thinkers," which are companies that actively understand and communicate the efforts that are needed in regards to the several social and environmental problems but are waiting for other agents to lead the way, and they will then follow. In the fourth category are the "doers," which are organizations that put their ideas into action, actively designing and applying efforts toward managing and overcoming the social and environmental crisis via a proactive approach.

In an earlier development, Welford (1995) established the ROAST scale, which has been used as an environmental performance scale with growing application in the evaluation of the environmental management system and performance of organizations. The scale is organized with a logic similar to that of the perspectives described previously, as oscillating along with a spectrum, developed via an incremental process, with increasing organizational commitment toward the social and environmental cause, presenting a growing engagement from the organization toward CSR. It helps in assessing the process of improving environmental performance, and for that aim, Welford (1995) argues that the model's essence is guided by the initial need to define an ultimate goal toward which the organization must move. The goal might not be achievable, but this by no means threatens the model's potential or even the organization's potential of success in improving its CSR. Nevertheless, inherent in the model's dynamic is that defining a goal is a prerequisite because that goal will serve as an upper boundary in guiding the organization's sustainable and competitive performance.

Hence "resistant organizations" represent the minimum organizational engagement in terms of environmental and social efforts; at the other extremity of this spectrum, the climax of organizational commitment toward the social and environmental topic can be observed, even though it reports an ideal and ultimate (nonetheless utopian) form of organization, labeled the "transcendent firm." In this the organization not only is fully aware of the social and environmental problems that the current global world faces, but also fully integrates those challenges as their priorities in terms of organizational performance, values, mission, strategy, policies, growth, and development, striving continually through their structure, actions, processes, and stakeholders to green its footprint to improve the overall social well-being and positively contribute to environmental conservation.

7.5 Toward integrated thinking: leadership and reporting as the pillars of sustainable organizations

7.5.1 Conceptualizing sustainable leadership

Based on the theoretical background described previously, in this section we advocate that leadership acts as a twofold catalyst and product of sustainable organizations. Hence using a circular logic, we postulate that sustainable organizations are both the result of and driven by a sustainable leadership model. Such an idea presents a refined approach to leadership, both in its theoretical domain and in its managerial models in the organizational and corporate milieu.

Following this line of thinking, the ground for such a review of the leadership approaches framed by the current worldwide sustainability motto acknowledges that while in the past the study of this organizational people management subject was focused on behavior, attributes, interactions, expectations, competencies, and styles among the several actors on the dyadic relationships (Hollenbeck, McCall, & Silzer, 2006), a fundamental evolution occurred as a result of the insights provided by various models. This involved reconfiguring approaches to take into account the concept of leadership in both its theoretical and managerial domains, which are deeply and intricately interconnected, as well as all the actors involved in the dynamic, resulting in values-oriented leadership theories (Chen & Li, 2013). Fry and Kriger (2009) systematize this renewed line of thinking by stressing the inadequacy and the insufficiency of the traditional approaches to leadership and suggesting a synthetized proposition claiming that a leadership process needs to go beyond a myopia approach of being only focused on "being," to indeed be complemented by emphasizing the vectors of "having" and "working," providing a more energetic, relational, and human approach to organizational people management.

Along the same lines but going further, McCauley, Van Velsor, and Ruderman (2010) point to this essential evolution in terms of leadership as a core people management approach in the organizational domain, suggesting that the vital step forward is rooted in the fact that the contemporary approaches conceive of leadership dynamics and results in the achievement of a collective, abandoning traditional and obsolete views that considered it as primarily the achievement of one leader and focused on the individual agent and his or her goals (Hambrick & Mason, 1984).

It is imperative that the contemporary leadership paradigms include the engagement of leaders with the various organizational actors in a broader scope, generating, developing, and capitalizing into win-win transactions, accessing and aiming at fulfilling the several wishes, goals, needs, and expectations of the myriad of organizational stakeholders (Sanford, 2011).

Based on the above framework, it is clear that companies, regardless of their industrial sector, are currently working with an increased sense of responsibility, awareness, and diligence toward the environment, society, and all their stakeholders (Cosby, 2014; Jackson & Seo, 2010; Kramar, 2014; Ren, Tang, & Jackson, 2020). More specifically, studies have suggested that for organizations that aspire to have effective and successful green practices in place, there is a need to enhance the individual green values of the employees. Hence the design and implementation of green practices in the organizational environment can be efficient and effective only when the individual green values are developed, assimilated, shared, and actualized by each organizational agent and, simultaneously, when it is required, as a core management goal of the organization, that those values be enhanced and enriched because the organization indeed acknowledges that such a line of thinking and action represents the chief outlook that can support organizations to achieve their ambitions in this scope (Edwards, 2010; Edwards, 2012; LiGraves & Sarkis, 2018; Liu, Mei, & Guo, 2020).

Empirical research on the topic suggests that the design and implementation of workplace policies, procedures, tools, and practices have a major impact on how employees think and behave in regard to environmental and social topics, boosting or hindering pro-environment behavior by the workforce that is strongly influenced by the leadership approach and by the leaders as individual agents themselves (Bhutto, Farooq, Talwar, Awan, & Dhir, 2021; Bissing-Olson, Iyer, Fielding, & Zacher, 2013; Dumont, Shen, & Deng, 2017; He, Morrison, & Zhang, 2021; Norton, Zacher, & Ashkanasy, 2014; Zhang, Ul-Durar, Akhtar, Zhang, & Lu, 2021), which will communicate the organization's vibrant and firm ambition and focus on developing its business on the way to greener entities (Mok, Shen, & Yang, 2018; Singh, Del Giudice, Chierici, & Graziano, 2020; Tan, Shuai, Shen, Hou, & Zhang, 2020).

Sustainable leadership involves a holistic approach to people management in the corporate environment, as it conceives of the organization as encompassing all of its agents and interactions, internal and external, including possible spillovers. Hence such an approach would be focusing well beyond the mere interaction of leaders with followers as a mutual exchange dyad in the form of transactional interactions. A review of the literature makes it clear that new models of leadership are emerging as a result of recent transformational perspectives in regards to leadership, which emphasize the symbolic behavior of leaders as a powerful stimulation toward the followers and focuses on feelings, direct relationships, personal image, wishes, and expectations, affecting their image of the leader, of themselves as organizational agents, and of the organization as a whole (Chen and Li, 2013). Vithessonthi (2009) recognizes that in a context in which there is an increased awareness of the social and ecological impacts of the broader set of organizational decisions, the leaders' attitudes toward sustainable development will greatly influence the decision-making process. This argument derives from the premise that postulates that individuals (organizational leaders, managers, and/or employees, etc.) have key roles in the fulfillment of organizational goals.

According to Peterlin et al. (2015), the journey toward the effective and efficient integration of the sustainability motto into the organizational environment requires a strategic decision-making process that integrates four main organizational areas via a joint complex dynamic: strategic sustainability, product and program sustainability, personnel sustainability, and financial sustainability. The first refers to the setting of a realistic vision and goals; the second is focused on the design and production of high-quality products, services, and programs; the third sheds light on the relevance of implementing practices, tools, procedures, and policies that provide an effective and efficient platform for workforce performance; and the fourth involves ensuring a financial reserve and implementing a contingency plan. Hence sustainable leadership approaches expand and enrich the organizational strategic decision-making process in the broader and interconnected global context of responsible leadership, a factor that assumes the clear advantage of this leadership approach, given that it marks an evolution from prior leadership theories. Such a benefit results from the fact that this contemporary leadership approach involves a clear and vivid orientation to serve, hence strongly directing strategies, policies, practices, tools, and procedures toward the increasing of overall social well-being and preservation of the natural environment, putting the common good ahead of self-interest. Such a leadership approach is also focused on the long-term interests of the myriad of organizational stakeholders, abandoning the traditional narrow focus on the dyad framed by the leader-follower interaction. Also worthy of mention, this reconfigured leadership perspective fully recognizes the onus of leaders and of leadership itself as bailiffs of the responsibility of organizational action, expansion, and development toward the enhancement of societal progress, safeguarding the common good and the development of a good quality of life not only for the present generations but also for future generations.

Such a leadership approach implies that sustainability connects to different and numerous aspects of organizational performance and development (Casserley & Critchley, 2010), this to say, it weights that top management needs to communicate, adapt, and implement a macro perspective of the organization (Avery & Bergsteiner, 2011). Hence via this leadership perspective will the community/civil society assume as a key stakeholder in the safeguard of greener practices toward the environment and the human population, given that it offers a natural and social expanded context within which the organizational function, purpose, and role are generated, developed, and upgraded, thus granting rich, wide-ranging, and contextualized decision-making processes, truly beholden of the social and ecological impacts. In other words, such an idea postulates that leaders need to ask both to themselves and to their stakeholders which systemic changes are they willing to design and to implement, and also via which policies, practices, tools, and procedures, so that clarity and firmness are observed in the attributes they conceive as the core basis to characterize the organizational sustainability. This to say, it is required, in such an enlarged and rich organizational decision-making process, that Sustainability evolves from a mere theoretical concept to represent a materialized managerial approach: to ensure that effectively and efficiently sustainability has sense in the organizational domain.

Therefore it can be advocated that sustainable leadership consists of a comprehensive systemic approach, since it is focused on several competing key factors that enable long-term development. First, sustainable leadership takes into consideration a wide range of complex interrelations among individuals, the business community, global markets, and the ecosystem, with the crucial understanding that an organization achieves success by respecting social values, utilizing a values-based strategic decision-making process, and protecting the natural environment, of which we all form an integral part (Kim, Kim, Han, Jackson, & Ployhart, 2017). Second, in strategic decision making, sustainable leadership involves acting responsibly. Avery and Bergsteiner (2011) claim that sustainable leadership builds communities, fosters collaboration among stakeholders, and promotes long-term value. The relevance of sustainable leadership for responsible strategic decision making in sustainable organizations is evident in the way it directs the attention of leaders toward four areas of consideration in making decisions. In other words, this refined contemporary leadership approach entails the strong, firm, and complex connection of four domains, which despite their idiosyncrasies, interact to form an open system: (1) the personal level, (2) the organizational level, (3) the social level, and (4) the ecological level. In the first domain, we focus on developing and ensuring personal psychological and physical health. In the second domain, attention is given to boosting the implementation of a work environment that is capable of inducing all organizational agents to develop, advance, and capitalize on a set of various skills and capabilities to foster organizational performance so that it becomes possible to achieve the organizational goals and growth milestones, ensuring a win-win situation for all stakeholders. In the third domain we consider the efforts as a whole from all agents, which form a cohesive but heterogeneous system that is known as civil society. In the fourth domain we take active and energetic steps toward the conservation of natural resources, hence imprinting a mindset directed toward sustainable environmental change.

On the basis of this background and a review of the literature, we can conclude that in terms of the theoretical framework of such a contemporary leadership approach, sustainable leadership is rooted in ethical leadership, as they share the vital vectors of public service, morality, and the common good (Brown & Treviño, 2006). Nonetheless, this approach also consists of a broader, richer, and more complex conceptual and managerial approach, given that its range of application includes the needs, goals, wishes, and aspirations of a wide assortment of organizational stakeholders, including future generations and the natural environment (Ahmad, Islam, Sadiq, & Kaleem, 2021).

The insights that we have provided so allow us to present a synthetic definition of the sustainable leadership concept. It can be clarified by its core vertices, which condense the essence and operational logic of the theoretical and managerial perspective of leadership as a vital people management approach in the organizational environment, hence postulating that sustainable leadership centers all the attention on accessing and nurturing the potential of future generations to enjoy an effective and dignified existence. Indeed, sustainable leadership assumes the leaders' firm, nonnegotiable, clear, and vivid dedication to holistically understanding the organizational system, conceiving of it as an open and complex system formed by a myriad of actors with often potentially conflicting interests but also acknowledging that all of them need to be in concert. This approach abandons the traditional focus of prior approaches that were based on the personal charisma or idealized influence of the leader/manager in persuading the followers. This concept can also emerge to explain the leadership model operating in a business environment that crystallizes the behaviors, attitudes, practices, policies, procedures, tools, and systems that create, develop, and capitalize on an enduring value for all the organizational stakeholders (Avery & Bergsteiner, 2011; Fernandez et al., 2020).

In another interesting perspective, which is nonetheless in communion with the previously discussed insights, Hargreaves (2007) suggests a definition of sustainable leadership as referring to an educational setting, given that, at its core, leadership generates, develops, and capitalizes via in-depth learning dynamics that generate and induce beneficial outcomes via spillover effects for all stakeholders in the present and future.

This line of reasoning is always directed toward the twofold starting point and endpoint, postulating that creating a sense of meaning enables and eases the effective and efficient commitment of stakeholders. Also, such a refined leadership approach is focused on the intellectual stimulation of stakeholders, an effort that develops into the active raising of awareness and boosts the motivation of all the organizational actors by inspiring them to action and boosting the greening of their values, imprinting on them a true need to serve the community (Avery & Bergsteiner, 2011; Avolio, Bass, & Jung, 1999; Bass, Avolio, Jung, & Berson, 2003).

As a corollary, we can postulate that all the presented proposals in regards to sustainable leadership emphasize that this refined approach to organizational people management (1) is exercised via win-win outcomes for a wide, complex, and diversified kaleidoscope of stakeholders; (2) adopts a long-term perspective, focusing on holistic models transcending a myopic concern with only the current state of affairs; (3) exercises leadership not only via the attitudes and behaviors of the leader and the corresponding influence via dyadic relationships with followers but also through other organizational systemic components, subsequently implementing a broader, richer and more complex organizational people management model; and (4) defines increased organizational value in terms of increased social common good. In essence, the concept of sustainable leadership provides a notable people management model in the contemporary and global organizational environment because it is distinguished by pursuing the value of sustainability at the joint connection of four vectors, namely, at the interconnection of the individual, organizational, social, and ecological levels, and addresses such efforts to both current and future generations.

By focusing on the insights that we have discussed, clear similarities and dissimilarities can be seen between this contemporary leadership approach and, for example, transformational, transactional, ethical, or even servant leadership. To conclude, it is important to highlight that sustainable leadership is still not, in terms of a conceptual and managerial approach, a mature theoretical concept as are other leadership approaches, traditional and contemporary. Moreover, according to Gurr (2007), it is both suitable and valuable to selectively elaborate a rigorous comparison between the sustainable leadership construct and other leadership theories and approaches as a fruitful starting point for the needed academic and managerial discussions.

7.5.2 Conceptualizing sustainable reporting

In the current globalized world, organizations are increasingly aware of the social and environmental problems and are therefore increasingly engaging in sustainability initiatives, reporting their activities and the corresponding spillover effects to a myriad of actors. Therefore we can postulate that the current developments in social accounting have emerged from the widespread recognition that the different players composing the organizational scenario have duties, goals, expectations, and responsibilities for a diverse assortment of needs and assets. This is vital to understand how performance should be accessed in organizational structures and how data can be combined to be reported at different scales and to diverse actors (Chesson, 2013).

In reviewing the literature with regard to the historical background of CSR and sustainability reporting, it becomes evident that the concerns about sustainability in corporate auditing and reporting have increased over time. Organizations are more often presenting organized and articulated responses to the growing calls for political structures, governmental guidelines, and national and international directives from nongovernmental agents, pressure groups, and other structured agents. The growing relevance of this topic, as well as its increasing presence in both academic debate and managerial organizational practices and policies, has occurred as a result of the widespread societal awareness and respective increased social scrutiny toward the environment and social causes (Haniffa & Cooke, 2005; Stiller & Daub, 2007). In a very interesting synthesis of the relevance of such discussions and as a creative means to alert the attention of stakeholders toward the need for sustainable reporting so that it becomes possible to achieve greener organizations, Spence (2009) advocated the relevance of CSR and sustainable reporting by an inverse logic, in comparison to that mentioned earlier. This author suggests that CSR consists of an expression of corporate egoism, having been purely and purposely designed to reinforce a comfortable corporate image and sense of self.

We can postulate that as a conceptual definition, sustainable reporting refers to a public report effort and product, produced by an organization, aimed at providing its internal and external stakeholders with a detailed overview of the organization's position and efforts in terms of the economic, environmental, and social dimensions of the overall organizational activity (Stiller and Daub, 2007). Within this context, organizations recognize that sustainability does not simply consist of a matter of good corporate citizenship; the exigency comes from the civil society, so sustainability becomes perceived as a central principle of intelligent management (Chen, Patten, & Roberts, 2008; del Mar Alonso-Almeida, Llach, & Marimon, 2014; Dobers & Springett, 2010; Savitz & Weber, 2006; van Staden & Hooks, 2007).

Reviewing the literature to understand the state of art in terms of this topic, we can see that the majority of the prior corporate sustainable reporting studies have been using ex post facto content analysis of annual reports and/or even other published data to study the relationship between the need and type of disclosure and their probable causal elements. In a summing-up effort, such an investigation trend resulted in the systematization of three groups of determinants (Adams, 2002). The first group refers to the corporate characteristics, highlighting industry group (Deegan & Gordon, 1996), company size (Hackston & Markus, 1996), and corporate age and risk (Roberts, 1992). The second group sheds light on the relevance of the general contextual factors, pointing out the organization's country of origin (Adams & Kuasirikun, 2000), its economic context (Guthrie & Parker, 1989), and cultural context (Haniffa & Cooke, 2005), the social and political context (Adams & Larrinaga-González, 2007; Burchell, Clubb, & Hopwood, 1985; Hogner, 1982; Vormedal & Ruud, 2009), and the ethical relativism the organization faces and acts upon (Lewis & Unerman, 1999), covering also the time and specific events (Burchell et al., 1985; Walden & Schwartz, 1997), the idiosyncrasies and the respective impact of the media coverage (Brown & Deegan, 1998), and the potential impact of the stakeholder power (Roberts, 1992). The third group of determinants reveals the importance of considering the contextual internal factors, namely, the composition of the corporate values, entailing as well the relevance of corporate governance (Archel, Husillos, Larrinaga, & Spence, 2009;

Beck, Campbell, & Shrives, 2010; Chen & Bouvain, 2009; Cowen, Ferreri, & Parker, 1987; Gibson & O'Donovan, 2007; Haniffa and Cooke, 2005).

Hence articulating the inputs that were presented in the preceding paragraphs, the literature exposes many theories that have been developed to explain corporate sustainable auditing and reporting, condensing them specifically into three main theoretical perspectives. In the first, we can highlight the relevance of legitimacy theory, given that it sustains the relevance of the social contract (Deegan, 2002) between the company and society, which is assumed as a preexisting condition so that organizations can ably, efficiently, and effectively continue their activities and develop and to expand them. In other words, according to this theoretical perspective, the planning and understanding of the organization's actions result only if the organization itself is primarily perceived by the wider society as having a legitimate position to act, influence, develop, and expand, hence adopting corporate auditing and reporting as a means to exert, maintain and reinforce its legitimacy (Buhr, 1998; De Villiers & Van Staden, 2006; Deegan & Rankin, 1997; Deegan, Cooper, & Shelly, 2006; Hogner, 1982). On the other hand, stakeholder theory sheds light on the evolving and complex relationship between the organization and the overall environment in which its actions take place (Gray, 2001, 2006), thus also covering the ability of organizations to assess and balance the conflicting and often incompatible demands and goals of their stakeholders (Roberts, 1992). Last but not least, institutional theory provides useful insights that help us to understand which mechanisms organizations use to align the design and the perceptions of their policies, practices, and characteristics with the social and cultural values of the society in which the organization acts. This theoretical construct also sheds light on how these values and their alignment, as above described, became institutionalized into the organization's structure and culture (Brignall & Modell, 2000; Covaleski, Dirsmith, & Samuel, 1996; Covaleski, Dirsmith, & White, 1987; Fogarty, 1996; Islam & Dellaportas, 2011; Rollins & Bremser, 1997).

A literature review on this topic postulates that even though institutional theory has been used as a recent perspective to understand CSR as well as providing useful inputs in terms of understanding sustainable auditing and reporting, the truth is that it complements both stakeholder and legitimacy theories in understanding how organizations generate, maintain, develop, and capitalize on legitimacy by recognizing and replying to the accelerating and changing social and institutional demands and expectations (Deegan & Unerman, 2006; Hossain, Momin, Rowe, & Quaddus, 2017; Rahaman, Lawrence, & Roper, 2004; Unerman & Bennett, 2004). In a very interesting contribution, Amran & Haniffa (2011) interpret sustainable reporting through institutional theory with the goal of ascertaining its possible determinants. This empirical research, based on the Malaysian context, indicates that all three mechanisms of isomorphism (i.e., the coercive, the normative, and the mimetic) contribute to the Malaysian companies' sustainability reporting. By using this framework, the authors suggest a model that postulates that the institutionalized activities emerge from the interrelated processes at the different, though deeply interconnected, individual, firm, and interfirm levels. The climax of such an idea is the position that even though these three components are different, they all connect through a complex web of interdependencies. Indeed, specifically for the case of sustainability reporting, this idea is rooted in the premise that the organizational culture, its respective working practices, and individual experiences and personal values affect not only the need but also the desirability of accessing performance, providing information, and reporting data to a variety of stakeholders.

Coercive isomorphism includes the pressure of national and international regulations, governmental action, and directives as well as the strength of the various socioeconomic and political institutions (Rahaman et al., 2004). This coercive stream derives from the political influence and the managerial scope of stakeholder theory, and in its dynamics it encompasses the process of organizations altering their institutional policies and practices as a consequence of both formal and informal pressures applied by stakeholders upon whom the organization depends. Mimetic isomorphism emerges from the premise that, given the uncertainties within the organizational environment, an organization will seek to emulate the policies, practices, procedures, and tools of other organizations because it perceives them to be more legitimate or successful, thus not only avoiding the risk of decreasing or even losing its legitimacy but also trying to maintain or even to enhance the competitive advantage (Unerman & Bennett, 2004). However, we can observe a connection with coercive isomorphism, as it is unlikely that organizations willingly embrace mechanisms to fulfill the demands and expectations of marginalized stakeholders unless the powerful stakeholders expect them to do so and see advantages in it (Unerman & Bennett, 2004). Then normative isomorphism emerges from group norms to cause the organization to adopt particular institutional practices or to fulfill certain professional expectations. In other words, it can be shown that this type of force derives from the phenomenon of professionalization, as it can be interpreted as the collective struggle of members of a given occupation to have the power and the responsibility to define the conditions, methods, tools, and practices to establish a cognitive base and a working platform for their professional kind (DiMaggio & Powell, 1983), thus also regulating the relationships of power in the job domain (Marshall & Brown, 2003).

In another very interesting contribution to understanding the roots and dynamics of sustainable auditing and reporting, Morimoto, Ash, and Hope reviewed the literature on green management, CSR, and corporate reporting to connect all the conceptual constructs. This effort resulted in a framework for social responsibility auditing, which assumes relevance as it is compatible with a commercially successful environmental audit system. Thus the authors present a prospective CSR audit protocol by making use of grounded theory as a means to overcome the tendency, clearly exposed in the literature, to perceive CSR as the social component of sustainable development, even though the literature still lacks agreement on its theoretical definition and managerial assessment. This prospective auditing and reporting protocol entails two main sections: (1) a description of the CSR system architecture, which covers board responsibility, corporate governance, stakeholder engagement, CSR policy, codes of conduct, complaints management, and environmental management standards and indexes, and (2) stakeholder factors, which are subdivided into six subsections, highlighting shareholders, clients and customers, suppliers, employees and contract staff, local inhabitants, and the general public.

According to Taylor (1992), acknowledging that green management consists of a continual process of assessment and improvement to create environmental and management excellence involves recognizing that a holistic approach to organizational management is needed to achieve such green goals. Hence a clear communion can be observed between this author and the previously mentioned authors, which we can synthesize as having been inspired by the earlier contribution of Taylor (1992), since all shed light on this need for a more ambitious analytical and managerial model in organizational practice so that sustainable development, in its overall significance, can be achieved.

Thus, connecting to the previous section of the chapter, we posit that together with sustainable leadership, implementing green management in the organizational domain implies efforts toward sustainable reporting. Indeed, organizational green management requires a rigorous and conscientious assessment of the state of the whole system, not just showing how companies merely comply with environmental legislation. In other words, sustainable reporting implies the need for competitive analysis of the organization's actions, policies, and structures, also contemplating the respective spillovers, based on the premise that such a holistic perspective provides the link between the audit tool itself, the reporting processes and results, and the configuration of the business environment, comprising also its social and environmental scope. This analysis is useful, reliable, and ambitious because it is not restricted to the internal functioning of the company. It assumes the ability to, as a precondition, scan the globe to assess competitors' actions, strategies, and ambitions and get informed on the updates of the normative directive and governmental guidelines as it, at the same time, focuses attention on assessment of the social and environmental pressures, thus being able to make predictions, addressing not only the trends but also the challenges and the opportunities (Achmad, Faisal, & Ulum, 2017; Aggarwal, 2013; Amran & Haniffa, 2011; Bae, Masud, Kaium, & Kim, 2018; Carter & Rogers, 2008; Chesson, 2013; Hahn & Kühnen, 2013; Kuzey & Uyar, 2017). It is also important to recognize that there are differences between the information requirements and data availability of both users and preparers of this sustainable reporting and auditing process to ensure the transparency, fairness, equity, and rigor on both the process, the mechanisms, the tools and, for sure, the outcomes, as Solomon and Lewis (2002) point out.

In regards to this, the literature shows that a key vehicle for enhancing a corporate image consists of the social report (Hess, 2007; Hess, Rogovsky, & Dunfee, 2002); hence it presents media pressure as a vivid rationale so that organizations compromise with sustainable reporting milestones. The truth is that organizations are increasingly committing to implementing nonmandatory environmental auditing, evidence of clear success in terms of a greener organizational mindset, which can be crystallized into Hess's (2007) argument postulating the critical need to establish an audit system that includes all aspects of a firm's performance, connecting its social, environmental, and business spheres. The core twofold motivation and outcome to engage in such efforts, from the organization'' side, derives from the value of the social report itself, given that it is perceived as synonymous with the creation of social transparency as well as institutionalizing responsible decision making and creative thinking in organizational management.

Nonetheless, as is clearly pointed out in the literature, the critical barrier to an efficient and effective sustainable reporting protocol resides not in the complexity and urgency of its creation, but rather in its implementation, based on the assumption that the strategy to assess and fulfill the goals of green management is formulated at all levels within the organization, thus balancing the often conflicting demands of a myriad of stakeholders. The vital challenge is not only to make it operational, but also to make it effective and efficient so that concrete outcomes can be achieved in the form of win-win solutions. Hence the key issue is to make the designed strategy work, because a strategy can be assumed to be successfully and correctly designed only if it is successfully and correctly implemented (Carter & Rogers, 2008).

Another vital link in the circuit of sustainable reporting comprises the communication flow, especially in regards to the clarity, objectivity, legitimacy, and strength of the message's content. Caution is needed in making use of the environment as a marketing tool by the organization, which can have a contradictory impact, especially when corporate structures are involved. Such a use will affect the organization's image and employee satisfaction, and, perhaps more relevant, it can help in providing a framework, together with other agents and actions, to reshape public policy (Carter & Rogers, 2008).

Hence by connecting and systematizing all the previously discussed inputs, alluding to the work of Morimoto, Ash, and Hope (2005), we can postulate key points to serve as pillars in constructing a model for sustainable reporting, which indeed help to stress the relevance of this vector as strategic in terms of the implementation of a green management mindset in the organizational domain. Hence such a sustainable reporting process or model should include significant stakeholder groups in the auditing process before assessing all the demands, needs, aspirations, and goals of these actors, balancing them via win-win dynamics in joint discussions and decision-making processes so that potentially conflicting demands can be articulated and fulfilled. Furthermore, the model should acknowledge the conceptual and theoretical difficulties that are still faced in conceiving discussions around sustainable development, environmental management, green management, organizational performance, CSR, and similar concepts.

The literature still does not convey a standard, objective, and consensually agreed definition of the previously discussed concepts. Therefore the respective theoretical constructs and managerial approaches multiply, and the debate is still an ongoing process of analyzing the different insights, culminating in the acknowledgment and embrace of diversity in individual perceptions of corporate performance and CSR in designing and implementing a sustainable reporting model. Hence the complexity of the subject may require specific training for auditors, managers, leaders, employees, board members, and so on (along with the organizational structure), given that, owing to this unpreparedness because of the still observed conceptual confusion and diversity, the assessment of some indices could be (dangerously) dependent on auditors' judgment and experience, potentially harming the accuracy of both the process and its results. In connection to this argument there is still the need to highlight the requirement of both quantitative and qualitative evaluation of corporate performance, never neglecting the essence of the triple bottom line approach in assessing corporate performance by the articulation of the social, environmental, and economic spheres in which the impact of the corporate actions is felt. This highlights the shortcomings of the "tick box" approach to auditing and reporting in terms of the organizational performance because implementing a green management model and mindset in the organizational structure, strategy, policies, and practices implies conceiving its actions in the present and future scenarios. In other words, a holistic perspective is needed that is oriented toward the long run so that the needs, expectations, and goals of the present generations can be attained without harming the potential to address and fulfill those of future generations. Therefore sustainable reporting should consist of an approach that goes beyond mere organizational compliance with laws, international directives, and normative dispositions.

As a climax and a synthesis effort, we can provide three practical tips for better sustainable reporting in the organizational domain so that in truth green management models can be applied in the current business organizational environment. The first tip addresses the urgency of ensuring that the organization's values fully and truly reflect its sustainability priorities because a circular logic can be established here, based on the idea that encouraging employees to commit to prosustainability behaviors requires a clear signal from the organization that is the employees are valued by the organizational structure and remain essential to the organization's present actions and future development. This argument is based on the premise that the current era, marked by the impacts of globalization, increases the need to make more salient the formal indicators in regards to the assessment and evaluation of organizational culture.

The second suggestion addresses the communication channels and flows, advocating the need for clear, objective, rigorous, strong, articulate, and cohesive messages whose content was previously filtered in the sense that it truly entails the balance of all the stakeholders' goals, demands, and expectations and is articulated via win-win solutions. This argument is rooted in the assumption that sustainable reporting, especially given its public dimension, allows organizations to communicate how they are contributing to sustainable development, not only to their key stakeholders but also to the overall society (Clark & Dickson, 2003; Parris & Kates, 2003a; 2003b). Perhaps one of the key purposes for organizations to adopt sustainable reporting is to use it as a public relations strategy, making it part of the organizational business strategy to gain and sustain competitive advantage. Because sustainable reporting can be used as a public relations strategy, concerns and efforts could then be addressed to the most influential agents in the organizational environment. Given that those agents hold significant resources for organizational growth, development, and expansion, such a strategy from the organization's side could increase its potential to successfully survive in the market.

It is also important to note that the communication should evolve in the sense that it needs to correspond (in its content, methods, and tools) with the overall scenario in terms of the changes in the key topics of sustainability for this specific case. Indeed, as organizational priorities relating to sustainable development progress and change through experience, so will their sustainable reporting needs, which means that the communication channels and tools also need to be updated, via a constant screening of the overall scenario, to ensure the credibility of the information.

The third guideline sheds light on the need to implement multidisciplinary approaches as a useful tool to condense a holistic perspective in regards to the design and implementation of a sustainable reporting model in the organization. Hence we suggest the usefulness of collaboration with behavioral accounting researchers, given that the current configurations of the worldwide business scenario are augmenting the need for synchronized discussions and actions so that as a result, the legitimacy, integrity, and reliability of sustainable reporting can be enhanced in terms of its key value for decision-making.

7.6 Conclusion

The active steps toward sustainable development depend on how organizations design and implement their strategies to deal with day-to-day activities; envisage their growth and expansion; and assess, manage, and capitalize on the spillovers of their actions. Hence imprinting sustainability effectively and efficiently in the organizational domain implies that organizations themselves need to commit to greener management models that are imprinted on the organizational structure, stakeholders, policies, practices, procedures, tools, metrics, and methods. Therefore a full and deep organizational engagement is needed toward the social and environmental causes, which is based on an embedded organizational awareness of the organizational role on the causes, consequences, and management of such issues.

Specifically, the chapter suggests that sustainable leadership and sustainable reporting are the key pillars to implement organizational sustainability.

In regards to sustainable leadership, the core rationale for it as a key pillar of organizational sustainability derives from the fact that in its very essence is the concept of values-based leadership in regards to both organizational and human development. This idea postulates that values-based approaches to leadership are vital to enhance organizational performance and development in regards to the evolution toward greener standards, as they facilitate the strategic decision-making process in corporate structures that are trying to embrace sustainability throughout their culture, structure, stakeholders, policies, processes, procedures, tools, and metrics. In particular, reviewing the literature, we can conclude that there are vital beneficial synergies that emerge as an outcome of the application of sustainable leadership models in the organizational domain, making it possible to integrate social and environmental topics and demands with economic ones in the current reinvigorated organizational decision-making process (Brown & Treviño, 2006; Viinamäki & 2009, 2012). To suggest the need for sustainable leadership so that sustainability can be efficiently and effectively implemented in the organizational structure means that organizations need to prioritize their leadership requirements as a key strategic decision, based on the premise that such a values-based orientation in terms of the understanding of leadership involves, as an organizational outcome, that those values-based leaders actively incorporate both stakeholder and organizational values into the organizational mindset, thus actively imprinting the awareness of the simultaneous social, environmental, and economic impacts of the organization's actions (Viinamäki, 2009, 2012). Hence the outcome consists of a rigorously selected decision that is profoundly informed by joint discussions and assessing and balancing potentially conflicting demands and expectations. The beneficial impacts of such a leadership approach can be expanded on the basis of the premise that those allied considerations generate stakeholder alignment, creating a positive psychological environment in the organization, in which pronatural and prosocial ideas can be expressed. Appropriate strategic decisions can thus designed and implemented with the support of all the organizational structure and actors. Such a beneficial outcome toward organizational sustainability can be achieved because this leadership approach includes sustainable people management practices, such as role-modeling, participatory communication, training, knowledge sharing, open and direct communication, mentoring, and reflection, profoundly acting on the organizational culture (Brown & Treviño, 2006; Cha & Edmondson, 2006; Graber & Kilpatrick, 2008; Prilleltensky, 2000; Schein, 2010; Viinamäki, 2009).

Sustainable leadership approaches are vital as the core of organizational sustainability, as they form around the idea that change occurs and evolves in concentric circles of integrative thinking within a given leadership network. Thus there is an ongoing and spillover dynamic in which the initiator of sustainable activity actively connects to other actors and inspires them to engage in and commit to the process of sustainable development, as they all cooperate in dynamics that, at their very end, crystallize into efforts directed at the common good by community building. This holistic component of sustainable leadership implies that an integrative mindset is required to implement organizational sustainability via leadership efforts, so it addresses, as a prerequisite for leadership, the necessity that leaders conceive of themselves and other human beings as parts of a much bigger system that forms a complex and intricate ecosystem resulting in synchronized needs, hence imprinting on leadership an ethic vector, which sustainability also clearly and vividly advocates (Maak & Pless, 2006).

Indeed, organizational green management requires a rigorous and conscientious assessment of the state of the whole system, not just how companies merely comply with environmental legislation. In other words, sustainable reporting is relevant to implement organizational sustainability as it involves the need for competitive analysis of the organization's actions, policies, and structures, also contemplating the respective spillovers, based on the premise that such a holistic viewpoint provides the link between the audit tool itself, the reporting processes and results, and the configuration of the business environment, comprising also its social and environmental scopes. This analysis is useful, reliable, and more ambitious because it is not restricted to the internal functioning of the company. It can, as a precondition, scan the globe to assess competitors' actions, strategies, and ambitions and become informed about the updates of the normative directive and governmental guidelines, as it at the same time focuses attention on the assessment of the social and environmental pressures, thus enabling the organization to make predictions, addressing not only trends but also challenges and opportunities (Achmad et al. 2017; Aggarwal, 2013; Amran & Haniffa, 2011; Bae et al. 2018; Carter & Rogers, 2008; Chesson, 2013; Hahn & Kühnen, 2013; Kuzey & Uyar, 2017).

This chapter has postulated that organizations need to focus on four crucial elements to achieve successful sustainable reporting. We highlighted the relevance of an effective and sustainable corporate leadership, and we focused attention on the importance of balanced stakeholder management (via joint discussion and win-win dynamics) and ensuring that conflicting demands are clarified, assessed, and fulfilled. It is also important that the potential impacts are understood and handled by all the organizational actors via a holistic paradigm and a widespread system. Another critical vector is a clear, firm, and vivid priority for CSR at the board level, thus clearly making public and active the organizational concern toward social and environmental causes so that such concerns are consciously integrated into corporate policy via ongoing actions with an increased commitment level in regards to the organizational environmental and social footprint. Another vital aspect to which organizations need to pay attention is how to boost active involvement and well-articulated coordination between the several actors and structures with governance and enforcement responsibilities and power, namely, governments, businesses, nongovernmental organizations, national and international cooperation platforms, and the civil society, so that regulation at the national and international level can be not only designed but also implemented and respected. Last but not least, another feature that is critical for sustainable reporting is a clear focus on the communication flow, especially in regards to the message's content, conveying its clarity, objectivity, legitimacy, and strength and acknowledging that caution is needed in making use of the environment as a marketing tool by organizations, which can have a contradictory impact when corporate structures are involved because it will affect the organizational image and employee satisfaction. Perhaps more relevant, it will help in providing a framework, together with other agents and actions, to reshape public policy (Carter & Rogers, 2008).

References

Achmad, T., Faisal, F., & Ulum, M. B. (2017). Nature of information provided and quality of sustainability reports assurance statements of fortune global 500 companies. *Journal of Applied Business Research, 33*(5), 1023–1034.

Adams, C. A. (2002). Internal organisational factors influencing corporate social and ethical reporting: Beyond current theorising. *Accounting, Auditing & Accountability Journal, 15* (2), 223–250.

Adams, C. A., & Kuasirikun, N. (2000). A comparative analysis of corporate reporting on ethical issues by UK and German chemical and pharmaceutical companies. *European Accounting Review, 9*(1), 53–79.

Adams, C. A., & Larrinaga-González, C. (2007). Engaging with organisations in pursuit of improved sustainability accounting and performance. *Accounting, Auditing & Accountability Journal, 20*(3), 333–355.

Aggarwal, P. (2013). Sustainability reporting and its impact on corporate financial performance: A literature review. *Indian Journal of Commerce and Management Studies, 4*(3), 51–59.

Ahmad, S., Islam, T., Sadiq, M., & Kaleem, A. (2021). Promoting green behavior through ethical leadership: A model of green human resource management and environmental knowledge. *Leadership & Organization Development Journal, 42*(4), 531–547.

Alfred, A. M., & Adam, R. F. (2009). Green management matters regardless. *Academy of Management Perspectives, 23*(3), 17–26.

Amran, A., & Haniffa, R. (2011). Evidence in development of sustainability reporting: A case of a developing country. *Business Strategy and the Environment, 20*(3), 141–156.

Archel, P., Husillos, J., Larrinaga, C., & Spence, C. (2009). Social disclosure, legitimacy theory and the role of the state. *Accounting, Auditing & Accountability Journal, 22*(8), 1284–1307.

Arguinis, H. (2011). Organizational responsibility: Doing good and doing well. In S. Zedeck (Ed.), *APA handbook of industrial and organizational psychology* (3, pp. 855–879). Washington, DC: American Psychological Association.

Avery, G. C., & Bergsteiner, H. (2011). Sustainable leadership practices for enhancing business resilience and performance. *Strategy & Leadership, 39*(3), 5–15.

Avolio, B. J., Bass, B. M., & Jung, D. I. (1999). Re-examining the components of transformational and transactional leadership using the multifactor leadership. *Journal of Occupational and Organizational Psychology, 72*(4), 441–462.

Azzone, G., Noci, G., Manzini, R., Welford, R., & Young, C. W. (1996). Defining environmental performance indicators: An integrated framework. *Business Strategy and the Environment, 5*(2), 69–80.

Babiak, K., & Trendafilova, S. (2011). CSR and environmental responsibility: Motives and pressures to adopt green management practices. *Corporate social responsibility and environmental management, 18*(1), 11–24.

Bae, S. M., Masud, M., Kaium, A., & Kim, J. D. (2018). A cross-country investigation of corporate governance and corporate sustainability disclosure: A signaling theory perspective. *Sustainability, 10*(8), 2611–2627.

Banerjee, S. B. (2001a). Corporate environmental strategies and actions. *Management Decision, 39*(1), 36–44.

Banerjee, S. B. (2001b). Managerial perceptions of corporate environmentalism: Interpretations from industry and strategic implications for organizations. *Journal of Management Studies, 38*(4), 489–513.

Bansal, P. (2003). From issues to actions: The importance of individual concerns and organizational values in responding to natural environmental issues. *Organization Science, 14*(5), 510–527.

Bass, B. M., Avolio, B. J., Jung, D. I., & Berson, Y. (2003). Predicting unit performance by assessing transformational and transactional leadership. *Journal of Applied Psychology, 88*(2), 207–218.

Beaumont, J. R., Whitaker, B., & Pedersen, L. (2013). *Managing the environment: Business opportunity and responsibility*. Oxford: Elsevier.

Beck, A. C., Campbell, D., & Shrives, P. J. (2010). Content analysis in environmental reporting research: Enrichment and rehearsal of the method in a British–German context. *The British Accounting Review, 42*(3), 207–222.

Benn, S., & Dunphy, D. (2013). *Corporate governance and sustainability: Challenges for theory and practice*. New York: Routledge.

Bennett, M., & James, P. (Eds.), (2017). *The Green bottom line: environmental accounting for management: current practice and future trends*. New York: Routledge.

Bhutto, T. A., Farooq, R., Talwar, S., Awan, U., & Dhir, A. (2021). Green inclusive leadership and green creativity in the tourism and hospitality sector: Serial mediation of green psychological climate and work engagement. *Journal of Sustainable Tourism, 29* (10), 1716–1737.

Bissing-Olson, M. J., Iyer, A., Fielding, K. S., & Zacher, H. (2013). Relationships between daily affect and pro-environmental behavior at work: The moderating role of pro-environmental attitude. *Journal of Organizational Behavior, 34*(2), 156–175.

Brignall, S., & Modell, S. (2000). An institutional perspective on performance measurement and management in the 'new public sector'. *Management Accounting Research, 11*(3), 281–306.

Brown, M. E., & Treviño, L. K. (2006). Ethical leadership: A review and future directions. *The Leadership Quarterly, 17*(6), 595–616.

Brown, N., & Deegan, C. (1998). The public disclosure of environmental performance information—A dual test of media agenda setting theory and legitimacy theory. *Accounting and Business Research, 29*, 21–41.

Buhr, N. (1998). Environmental performance, legislation and annual report disclosure: The case of acid rain and Falconbridge. *Accounting, Auditing & Accountability Journal, 11*(2), 163–190.

Burchell, S., Clubb, C., & Hopwood, A. (1985). Accounting in its social context: Towards a history of value added in the United Kingdom. *Accounting, Organizations and Society, 10* (4), 381–413.

Cabezas, H., & Faith, B. D. (2002). Towards a theory of sustainable systems. *Fluid Phase Equilibria, 194*, 3–14.

Cairncross, F. (2014). *Green Inc.: Guide to business and the environment*. New York: Routledge.

Carter, C. R., & Rogers, D. S. (2008). A framework of sustainable supply chain management: Moving toward new theory. *International Journal of Physical Distribution & Logistics Management, 38*(5), 360–387.

Casserley, T., & Critchley, B. (2010). A new paradigm of leadership development. *Industrial and Commercial Training, 42*(6), 287–295.

Cha, S. E., & Edmondson, A. C. (2006). When values backfire: Leadership, attribution, and disenchantment in a values-driven organization. *The Leadership Quarterly, 17*(1), 57–78.

Chen, C. Y., & Li, C. I. (2013). Assessing the spiritual leadership effectiveness: The contribution of follower's self–concept and preliminary tests for moderation of culture and managerial position. *The Leadership Quarterly, 24*(1), 240–255.

Chen, J. C., Patten, D. M., & Roberts, R. W. (2008). Corporate charitable contributions: A corporate social performance or legitimacy strategy? *Journal of Business Ethics, 82*(1), 131–144.

Chen, S., & Bouvain, P. (2009). Is corporate responsibility converging? A comparison of corporate responsibility reporting in the USA, UK, Australia, and Germany. *Journal of Business Ethics, 87*(1), 299–317.

Chesson, J. (2013). Sustainable development: Connecting practice with theory. *Journal of Environmental Assessment Policy and Management, 15*(1), 1–27.

Clark, W. C., & Dickson, N. M. (2003). Sustainability science: The emerging research program. *Proceedings of the National Academy of Sciences, 100*(14), 8059–8061.

Coglianese, C., & Nash, J. (2010). *Environmental management systems and the new policy agenda. Regulating from the inside*. New York: Routledge.

Cohen, B., & Winn, M. I. (2007). Market imperfections, opportunity and sustainable entrepreneurship. *Journal of Business Venturing, 22*(1), 29–49.

Collins, J. (2001). *Good to great: Why some companies make the leap, and others don't*. New York: Harper Business.

Collins, J. C., & Porras, J. I. (2000). *Built to last: Successful habits of visionary companies*. New York: Harper Collins.

Cosby, D. M. (2014). Sustainability program leadership for human resource development professionals: A competency model. *Journal of Organizational Culture, Communications and Conflict, 18*(2), 79–86.

Costello, M. W. (2008). 13 steps to 'green' your business. *Business & Economic Review, 54*(4), 6–9.

Covaleski, M. A., Dirsmith, M. W., & Samuel, S. (1996). Managerial accounting research: The contributions of organizational and sociological theories. *Journal of Management Accounting Research, 8*, 1–36.

Covaleski, M. A., Dirsmith, M. W., & White, C. E. (1987). Economie consequences: The relationship between financial reporting and strategic planning, management and operating control decisions. *Contemporary Accounting Research, 3*(2), 408–429.

Cowen, S. S., Ferreri, L. B., & Parker, L. D. (1987). The impact of corporate characteristics on social responsibility disclosure: A typology and frequency based analysis. *Accounting Organization and Society, 12*(2), 11–122.

Crane, A., McWilliams, A., Matten, D., Moon, J., & Siegel, D. S. (Eds.), (2008). *The Oxford handbook of corporate social responsibility*. Oxford: Oxford University Press.

Daly, H. E. (1991). Elements of environmental macroeconomics. In R. Costanza (Ed.), *Ecological economics: The science and management of sustainability* (pp. 32–46). New York: Columbia University Press.

Darnall, N., Jolley, G. J., & Handfield, R. (2008). Environmental management systems and green supply chain management: Complements for sustainability? *Business Strategy and the Environment, 17*(1), 30–45.

De Villiers, C., & Van Staden, C. J. (2006). Can less environmental disclosure have a legitimising effect? Evidence from Africa. *Accounting, Organizations and Society, 31*(8), 763–781.

Dean, T. J., & McMullen, J. S. (2007). Toward a theory of sustainable entrepreneurship: Reducing environmental degradation through entrepreneurial action. *Journal of Business Venturing, 22*(1), 50–76.

Deegan, C. (2002). Introduction: The legitimising effect of social and environmental disclosures–a theoretical foundation. *Accounting, Auditing & Accountability Journal, 15*(3), 282–311.

Deegan, C., & Gordon, B. (1996). A study of the environmental disclosure practices of Australian corporations. *Accounting and Business Research, 26*(3), 187–199.

Deegan, C., & Rankin, M. (1997). The materiality of environmental information to users of annual reports. *Accounting, Auditing & Accountability Journal, 10*(4), 562–583.

Deegan, C., & Unerman, J. (2006). *Financial accounting theory: European edition*. Maidenhead: McGraw-Hill.

Deegan, C., Cooper, B. J., & Shelly, M. (2006). An investigation of TBL report assurance statements: UK and European evidence. *Managerial Auditing Journal, 21*(4), 329–371.

del Mar Alonso-Almeida, M., Llach, J., & Marimon, F. (2014). A closer look at the 'Global Reporting Initiative'sustainability reporting as a tool to implement environmental and social policies: A worldwide sector analysis. *Corporate Social Responsibility and Environmental Management, 21*(6), 318–335.

Delmas, M., & Toffel, M. W. (2004). Stakeholders and environmental management practices: an institutional framework. *Business Strategy and the Environment, 13*(4), 209–222.

DiMaggio, P. J., & Powell, W. W. (1983). The iron cage revisited: Institutional isomorphism and collective rationality in organizational fields. *American Sociology Review, 48*, 146–160.

Dobers, P., & Springett, D. (2010). Corporate social responsibility: Discourse, narratives and communication. *Corporate Social Responsibility and Environmental Management, 17*(2), 63–69.

Dovers, S. R., & Handmer, J. W. (1993). Contradictions in sustainability. *Environmental Conservation, 20*(3), 217–222.

Dresner, S. (2012). *The principles of sustainability*. New York: Routledge.

Drucker, P. (2001). *The Essential Drucker*. Oxford: Butterworth Heinemann.

Dumont, J., Shen, J., & Deng, X. (2017). Effects of green HRM practices on employee workplace green behavior: The role of psychological green climate and employee green values. *Human Resource Management, 56*(4), 613–627.

Edwards, A. R. (2010). *Thriving beyond sustainability: Pathways to a resilient society*. Canada: New Society Publishers.

Edwards, K. E. (2012). Moving beyond green: Sustainable development toward healthy environments, social justice, and strong economies. *New Directions for Student Services, 2012*(137), 19–28.

Fernandes, D. S., & Machado, C. F. (2020a). How to look at organizations and human resource management in the economy of the future? In C. e Machado, & J. P. Davim (Eds.), *Circular economy and engineering a new ecologically efficient model* (pp. 89–117). Switzerland: Springer.

Fernandes, D. S., & Machado, C. F. (2020b). Stop camouflaging it in green: Do not confuse corporate social responsibility with sustainable management. In C. Machado, & J. P. Davim (Eds.), *Sustainable management for managers and engineers* (pp. 225–250). London: John Wiley & Sons.

Fernandez, A., Kullu, F. D., & Shankar, R. (2020). *A grounded research approach to sustainable leadership practices and competencies. Sustainable human resource management* (pp. 71–86). Singapore: Springer.

Florea, L., Cheung, Y. H., & Herndon, N. C. (2013). For all good reasons: Role of values in organizational sustainability. *Journal of Business Ethics, 114*(3), 393–408.

Florida, R., Atlas, M., & Cline, M. (2001). What makes companies green? Organizational and geographic factors in the adoption of environmental practices. *Economic Geography, 77*(3), 209–224.

Fogarty, T. J. (1996). The imagery and reality of peer review in the US: Insights from institutional theory. *Accounting, Organizations and Society, 21*(2/3), 243–267.

Foote, J., Gaffney, N., & Evans, J. R. (2010). Corporate social responsibility: Implications for performance excellence. *Total Quality Management, 21*(8), 799–812.

Freeman, P. K., & Kunreuther, H. (2002). Environmental risk management for developing countries. *The Geneva Papers on Risk and Insurance-Issues and Practice, 27*(2), 196–214.
Freeman, R. E. (1984). *Strategic management: A stakeholder approach.* Boston: Pitman.
Freeman, R. E. (2010). Managing for stakeholders: Trade-offs or value chain. *Journal of Business Ethics, 96*, 7–9.
Fry, L., & Kriger, M. (2009). Towards a theory of being-centered leadership: Multiple levels of being as context for effective leadership. *Human Relations, 62*(11), 1667–1696.
Galeazzo, A., Furlan, A., & Vinelli, A. (2014). Lean and green in action: Interdependencies and performance of pollution prevention projects. *Journal of Cleaner Production, 85*, 191–200.
Galpin, T., & Whittington, J. L. (2012). Sustainability leadership: From strategy to results. *Journal of Business Strategy, 33*(4), 40–48.
Garzella, S., & Fiorentino, R. (2014). An integrated framework to support the process of green management adoption. *Business Process Management Journal, 20*(1), 68–89.
Ghobadian, A., Viney, H., Liu, J., & James, P. (1998). Extending linear approaches to mapping corporate environmental behaviour. *Business Strategy and the Environment, 7*(1), 13–23.
Gibson, K., & O'Donovan, G. (2007). Corporate governance and environmental reporting: An Australian study. *Corporate Governance: An International Review, 15*(5), 944–956.
Goodland, R., & Daly, H. (1996). Environmental sustainability: Universal and non-negotiable. *Ecological Applications, 6*(4), 1002–1017.
Graber, D. R., & Kilpatrick, A. O. (2008). Establishing values–based leadership and value systems in healthcare organizations. *Journal of Health and Human Services Administration, 31*(2), 179–197.
Graves, L. M., & Sarkis, J. (2018). The role of employees' leadership perceptions, values, and motivation in employees' provenvironmental behaviors. *Journal of Cleaner Production, 196*, 576–587.
Gray, R. (2001). Thirty years of social accounting, reporting and auditing: What (if anything) have we learnt? *Business Ethics: A European Review, 10*(1), 9–15.
Gray, R. (2006). Social, environmental and sustainability reporting and organisational value creation? Whose value? Whose creation? *Accounting, Auditing & Accountability Journal, 19*(6), 793–819.
Griffiths, A., Dunphy, D., & Benn, S. (2003). Building corporate sustainability. *Training and Development in Australia, 30*(2), 20–22.
Gurr, D. (2007). Developing sustainable leadership. *Australian Journal of Education, 51*(3), 341–343.
Guthrie, J., & Parker, L. D. (1989). Corporate social disclosure practice: A comparative international analysis. *Advances in Public Interest Accounting, 3*, 159–175.
Hackston, D., & Markus, J. M. (1996). Some determinants of social and environmental disclosures in New Zealand companies. *Accounting, Auditing & Accountability Journal, 9*(1), 77–108.
Haden, S. P., Oyler, J. D., & Humphreys, J. H. (2009). Historical, practical, and theoretical perspectives on green management: An exploratory analysis. *Management Decision, 47* (7), 1041–1055.
Hahn, R., & Kühnen, M. (2013). Determinants of sustainability reporting: A review of results, trends, theory, and opportunities in an expanding field of research. *Journal of Cleaner Production, 59*, 5–21.
Haldar, S. (2019). Green entrepreneurship in theory and practice: Insights from India. *International Journal of Green Economics, 13*(2), 99–119.
Hambrick, D. C., & Mason, P. A. (1984). Upper echelons: The organization as a reflection of its top managers. *Academy of Management Review, 9*(2), 193–206.
Haniffa, R. M., & Cooke, T. E. (2005). The impact of culture and governance on corporate social reporting. *Journal of Accounting & Public Policy, 24*(5), 391–430.

Hardy, A., Beeton, R. J., & Pearson, L. (2002). Sustainable tourism: An overview of the concept and its position in relation to conceptualisations of tourism. *Journal of Sustainable Tourism, 10*(6), 475–496.

Hargreaves, A. (2007). Sustainable leadership and development in education: Creating the future, conserving the past. *European Journal of Education, 42*(2), 223–233.

Harris, L. C., & Crane, A. (2002). The greening of organizational culture: Management views on the depth, degree and diffusion of change. *Journal of Organizational Change Management, 15*(3), 214–234.

Hart, S. L. (1995). A natural-resource-based view of the firm. *Academy of Management Review, 20*(4), 986–1014.

Hart, S. L., & Ahuja, G. (1996). Does it pay to be green? An empirical examination of the relationship between emission reduction and firm performance. *Business strategy and the Environment, 5*(1), 30–37.

Hass, J. L. (1996). Environmental ('green') management typologies: An evaluation, operationalization and empirical development. *Business Strategy and the Environment, 5*(2), 59–68.

Hawken, P. (1993). *The ecology of commerce: A declaration of sustainability*. New York: Collins Business.

He, J., Morrison, A. M., & Zhang, H. (2021). Being sustainable: The three-way interactive effects of CSR, green human resource management, and responsible leadership on employee green behavior and task performance. *Corporate Social Responsibility and Environmental Management, 28*(3), 1043–1054.

Hervani, A. A., Helms, M. M., & Sarkis, J. (2005). Performance measurement for green supply chain management. *Benchmarking: An International Journal, 12*(4), 330–353.

Hess, D. (2007). Social reporting and new governance regulation: The prospects of achieving corporate accountability through transparency. *Business Ethics Quarterly, 17*(3), 453–476.

Hess, D., Rogovsky, N., & Dunfee, T. W. (2002). The next wave of corporate community involvement: Corporate social initiatives. *California Management Review, 44*(2), 110–125.

Hogner, R. H. (1982). Corporate social reporting: Eight decades of development at US Steel. *Research in Corporate Performance and Policy, 4*(1), 243–250.

Hollenbeck, G. P., McCall, M. W., Jr, & Silzer, R. F. (2006). Leadership competency models. *The Leadership Quarterly, 17*(4), 398–413.

Hörisch, J., Freeman, R. E., & Schaltegger, S. (2014). Applying stakeholder theory in sustainability management: Links, similarities, dissimilarities, and a conceptual framework. *Organization & Environment, 27*(4), 328–346.

Hossain, M. M., Momin, M. A., Rowe, A. L., & Quaddus, M. (2017). Corporate social and environmental reporting practices: A case of listed companies in Bangladesh. *Sustainability Accounting, Management and Policy Journal, 8*(2), 138–165.

Islam, M., & Dellaportas, S. (2011). Perceptions of corporate social and environmental accounting and reporting practices from accountants in Bangladesh. *Social Responsibility Journal, 7*(4), 649–664.

Jackson, S. E., & Seo, J. (2010). The greening of strategic HRM scholarship. *Organization Management Journal, 7*(4), 278–290.

James, P. (1994). Business environmental performance measurement. *Business Strategy and the Environment, 3*(2), 59–67.

Jennings, D. F., & Lumpkin, J. R. (1992). Insights between environmental scanning activities and Porter's generic strategies: An empirical analysis. *Journal of Management, 18*(4), 791–803.

Johnson, D. B. (1998). Green businesses: Perspectives from management and business ethics. *Society & Natural Resources, 11*(3), 259–266.

Kautto, P. (2006). New instruments–old practices? The implications of environmental management systems and extended producer responsibility for design for the environment. *Business Strategy and the Environment, 15*(6), 377–388.

Ketola, T. (1993). The seven sisters: Snow Whites, dwarfs or evil queens? A comparison of the official environmental policies of the largest oil corporations in the world. *Business Strategy and the Environment, 2*(3), 22–33.

Ketola, T. (1997). A map of Neverland: The role of policy in strategic environmental management. *Business Strategy and the Environment, 6*(1), 18–33.

Ketola, T. (2007). Ten years later: Where is our common future now? *Business Strategy and the Environment, 16*(3), 171–189.

Ketola, T. (2008). A holistic corporate responsibility model: Integrating values, discourses and actions. *Journal of Business Ethics, 80*(3), 419–435.

Kiewiet, D. J., & Vos, J. F. (2007). Organisational sustainability: A case for formulating a tailor-made definition. *Journal of Environmental Assessment Policy and Management, 9*(01), 1–18.

Kim, A., Kim, Y., Han, K., Jackson, S. E., & Ployhart, R. E. (2017). Multilevel influences on voluntary workplace green behavior: Individual differences, leader behavior, and coworker advocacy. *Journal of Management, 43*(5), 1335–1358.

King, A. A., & Lenox, M. J. (2001). Does it really pay to be green? An empirical study of firm environmental and financial performance. *Journal of Industrial Ecology, 5*(1), 105–116.

Kivimaa, P., & Kautto, P. (2010). Making or breaking environmental innovation? Technological change and innovation markets in the pulp and paper industry. *Management Research Review, 33*(4), 289–305.

Koe, W. L., & Majid, I. A. (2014). Socio-cultural factors and intention towards sustainable entrepreneurship. *Eurasian Journal of Business and Economics, 7*(13), 145–156.

Kramar, R. (2014). Beyond strategic human resource management: Is sustainable human resource management the next approach? *The International Journal of Human Resource Management, 25*(8), 1069–1089.

Kuzey, C., & Uyar, A. (2017). Determinants of sustainability reporting and its impact on firm value: Evidence from the emerging market of Turkey. *Journal of Cleaner Production, 143*, 27–39.

Lamond, D. (2009). Why and how green management matters. *Management Decision, 47*(7), 1033–1037.

Lee, K. H., & Ball, R. (2006). *Achieving sustainable corporate competitiveness. Managing the Business Case for Sustainability: The Integration of Social. Environmental and Economic Performance*. New York: Taylor & Francis.

Lefebvre, É., Lefebvre, L. A., & Talbot, S. (2003). Determinants and impacts of environmental performance in SMEs. *R&D Management, 33*(3), 263–283.

Lewis, L., & Unerman, J. (1999). Ethical relativism: A reason for differences in corporate social reporting? *Critical Perspectives on Accounting, 10*(4), 521–547.

Liu, Z., Mei, S., & Guo, Y. (2020). Green human resource management, green organization identity and organizational citizenship behavior for the environment: The moderating effect of environmental values. *Chinese Management Studies, 15*(2), 290–304.

Loknath, Y., Azeem, B.A. (2017). Green management – Concept and strategies. In: *Proceedings of the National Conference on Marketing and Sustainable Development*, 13–14 October 2017 (vol. 13, pp. 688–702). New Delhi, India. <http://www.aimsinternational.org/myconference/cd/PDF/MSD4–6111-Done.pdf> Accessed 29.10.22.

Maak, T., & Pless, N. M. (2006). Responsible leadership in a stakeholder society – a relational perspective. *Journal of Business Ethics, 66*(1), 99–115.

Makower, J. (1994). *The E Factor: The Bottom Line Approach to Environmentally Responsible Business*. New York: Penguin Books.

Marshall, R. S., & Brown, D. (2003). Corporate environmental reporting: What's in a metric? *Business Strategy and the Environment, 12*(2), 87–106.
Matthews, J., & Senyard, J. (2010). Characteristics of sustainable entrepreneurship: Some early explorations from the CAUSEE study. In: *Proceedings of the 7th AGSE International Entrepreneurship Research Exchange* (pp. 903–917).
McCauley, C. D., Van Velsor, E., & Ruderman, M. N. (2010). Introduction: Our view of leadership development. In E. Van Velsor, C. D. McCauley, & M. N. Ruderman (Eds.), *The center for creative leadership handbook of leadership development* (3rd ed., pp. 1–26). San Francisco, CA: Jossey-Bass.
McWilliams, A., & Siegel, D. S. (2011). Creating and capturing value: Strategic corporate social responsibility, resource-based theory, and sustainable competitive advantage. *Journal of Management, 37*(5), 1480–1495.
McWilliams, A., Siegel, D. S., & Wright, P. M. (2006). Corporate social responsibility: Strategic implications. *Journal of Management Studies, 43*(1), 1–18.
Mirvis, P., & Googins, B. (2006). Stages of corporate citizenship. *California Management Review, 48*(2), 104–126.
Mohrman, S. A., & Worley, C. G. (2010). The organizational sustainability journey: Introduction to the special issue. *Organizational Dynamics, 39*(4), 289–294.
Mok, K. Y., Shen, G. Q., & Yang, R. (2018). Stakeholder complexity in large scale green building projects: A holistic analysis towards a better understanding. *Engineering, Construction and Architectural Management, 25*(11), 1454–1474.
Morimoto, R., Ash, J., & Hope, C. (2005). Corporate social responsibility audit: From theory to practice. *Journal of Business Ethics, 62*(4), 315–325.
Muñoz, P., Janssen, F., Nicolopoulou, K., & Hockerts, K. (2018). Advancing sustainable entrepreneurship through substantive research. *International Journal of Entrepreneurial Behavior & Research, 24*(2), 322–332.
Netherwood, A. (1996). Environmental management systems. *Corporate environmental Management 1: Systems and Strategies, 2*, 35–58.
Norton, T. A., Zacher, H., & Ashkanasy, N. M. (2014). Organisational sustainability policies and employee green behaviour: The mediating role of work climate perceptions. *Journal of Environmental Psychology, 38*, 49–54.
Orlitzky, M., Siegel, D. S., & Waldman, D. A. (2011). Strategic corporate social responsibility and environmental sustainability. *Business & Society, 50*(1), 6–27.
Parris, T. M., & Kates, R. W. (2003a). Characterizing and measuring sustainable development. *Annual Review of Environment and Resources, 28*(1), 559–586.
Parris, T. M., & Kates, R. W. (2003b). Characterizing a sustainability transition: Goals, targets, trends, and driving forces. *Proceedings of the National Academy of Sciences, 100*(14), 8068–8073.
Peterlin, J., Pearse, N., & Dimovski, V. (2015). Strategic decision making for organizational sustainability: The implications of servant leadership and sustainable leadership approaches. *Economic and Business Review, 17*(3), 273–290.
Porter, M. E., & Van der Linde, C. (1995). Toward a new conception of the environment-competitiveness relationship. *Journal of Economic Perspectives, 9*(4), 97–118.
Prilleltensky, I. (2000). Value-based leadership in organizations: Balancing values, interests, and power among citizens, workers, and leaders. *Ethics & Behavior, 10*(2), 139–158.
Puopolo, G. W., Teti, E., & Milani, V. (2015). Does the market reward for going green? *Journal of Management Development, 34*(6), 729–742.
Rahaman, A. S., Lawrence, S., & Roper, J. (2004). Social and environmental reporting at the VRA: Institutionalised legitimacy or legitimation crisis? *Critical perspectives on Accounting, 15*(1), 35–56.
Rajasekaran, B. (2013). Sustainable entrepreneurship: Past researches and future directions. *Journal of Entrepreneurship and Management, 2*(1), 20–27.

Raufflet, E., Bres, L., & Filion, L. J. (2014). Desenvolvimento sustentável e empreendedorismo. *Iberoamerican Journal of Entrepreneurship and Small Business, 3*(1), 3–32.

Ren, S., Tang, G., & Jackson, S. E. (2020). Effects of Green HRM and CEO ethical leadership on organizations' environmental performance. *International Journal of Manpower, 42*(6), 961–983.

Ricart, J. E., Rodríguez, M. Á., & Sanchez, P. (2005). Sustainability in the boardroom: An empirical examination of Dow Jones sustainability world index leaders. *Corporate Governance: the International Journal of Business in Society, 5*(3), 24–41.

Roberts, R. W. (1992). Determinants of corporate social responsibility disclosure. *Accounting, Organizations and Society, 17*(6), 595–612.

Rollins, T. P., & Bremser, W. G. (1997). The SEC's enforcement actions against auditors: An auditor reputation and institutional theory perspective. *Critical Perspectives on Accounting, 8*, 191–206.

Roome, N. (1992). Developing environmental management strategies. *Business Strategy and the Environment, 1*(1), 11–24.

Rothenberg, S., Maxwell, J., & Marcus, D. A. (1992). Issues in the implementation of proactive environmental strategies. *Business Strategy and the Environment, 1*(4), 1–12.

Royal, C., Daneshgar, F., & O'Donnell, L. (2003). Facilitating organisational sustainability through expert investment systems. *Electronic Journal of Knowledge Management, 1*(2), 167–176.

Sanford, C. (2011). *The responsible business: Reimagining sustainability and success*. New York: John Wiley & Sons.

Savitz, A. W., & Weber, K. (2006). *The triple bottom line: How today's best run companies are achieving economic. Social and environmental success and how you can too*. New York: John Wiley & Sons.

Schein, E. (2010). *Organizational culture and leadership* (4th ed). San Francisco, CA: Jossey-Bass.

Senge, P. M., Smith, B., Kruschwitz, N., Laur, J., & Schley, S. (2008). *The necessary revolution: How individuals and organizations are working together to create a sustainable world*. New York: Doubleday.

Sexton, K., Marcus, A. A., Easter, K. W., & Burkhardt, T. D. (Eds.), (1999). *Better environmental decisions: Strategies for governments, businesses, and communities*. Washington, DC: Island Press.

Sharma, R. R. (2020). Green management and circular economy for sustainable development. *Vision: The Journal of Business Perspective, 24*(1), 7–8.

Sharma, S. (2000). Managerial interpretations and organizational context as predictors of corporate choice of environmental strategy. *Academy of Management Journal, 43*(4), 681–697.

Sharma, S., & Ruud, A. (2003). On the path to sustainability: Integrating social dimensions into the research and practice of environmental management. *Special Issue: Corporate Social Responsibility & Governance for Sustainability, 12*(4), 205–214.

Sharma, S., & Vredenburg, H. (1998). Proactive corporate environmental strategy and the development of competitively valuable organizational capabilities. *Strategic Management Journal, 19*(8), 729–753.

Siegel, D. S. (2009). Green management matters only if it yields more green: An economic/strategic perspective. *Academy of Management Perspectives, 23*(3), 5–16.

Simpson, A. (1991). *The greening of global investment – how the environment. Ethics and politics are reshaping strategies*. London: The Economist Publications.

Singh, S. K., Del Giudice, M., Chierici, R., & Graziano, D. (2020). Green innovation and environmental performance: The role of green transformational leadership and green human resource management. *Technological Forecasting and Social Change, 150*, 119–762.

Slaper, T. F., & Hall, T. J. (2011). The triple bottom line: What is it and how does it work. *Indiana Business Review, 86*(1), 4–8.

Smith, P. A. (2012). The importance of organizational learning for organizational sustainability. *The Learning Organization, 19*(1), 4–10.
Solomon, A., & Lewis, L. (2002). Incentives and disincentives for corporate environmental disclosure. *Business Strategy and the Environment, 11*(3), 154–169.
Spence, C. (2009). Social and environmental reporting and the corporate ego. *Business Strategy and the Environment, 18*(4), 254–265.
Stainer, A., & Stainer, L. (1997). Ethical dimensions of environmental management. *European Business Review, 97*(5), 224–230.
Starik, M., & Marcus, A. A. (2000). Introduction to the special research forum on the management of organizations in the natural environment: A field emerging from multiple paths, with many challenges ahead. *Academy of Management Journal, 43*(4), 539–547.
Stead, J. G., & Stead, E. (2000). Eco-enterprise strategy: Standing for sustainability. *Journal of Business Ethics, 24*(4), 313–329.
Steger, U. (1996). Managerial issues in closing the loop. *Business Strategy and the Environment, 5*(4), 252–268.
Stiller, Y., & Daub, C. H. (2007). Paving the way for sustainability communication: Evidence from a Swiss study. *Business Strategy and the Environment, 16*(7), 474–486.
Tan, Y., Shuai, C., Shen, L., Hou, L., & Zhang, G. (2020). A study of sustainable practices in the sustainability leadership of international contractors. *Sustainable Development, 28* (4), 697–710.
Taylor, N., Barker, K., & Simpson, M. (2003). Achieving 'sustainable business': A study of perceptions of environmental best practice by SMEs in South Yorkshire. *Environment and Planning C: Government and Policy, 21*(1), 89–105.
Taylor, S. R. (1992). Green management: The next competitive weapon. *Futures, 24*(7), 669–680.
Thomas, T. E., & Lamm, E. (2012). Legitimacy and organizational sustainability. *Journal of Business Ethics, 110*(2), 191–203.
Unerman, J., & Bennett, M. (2004). Increased stakeholder dialogue and the internet: Towards greater corporate accountability or reinforcing capitalist hegemony? *Accounting, Organizations and Society, 29*(7), 685–707.
van Staden, C. J., & Hooks, J. (2007). A comprehensive comparison of corporate environmental reporting and responsiveness. *The British Accounting Review, 39*(3), 197–210.
Viinamäki, O. P. (2009). Intra-organizational challenges of values-based leadership. *Electronic Journal of Business Ethics and Organization Studies, 14*(2), 6–13.
Viinamäki, O. P. (2012). Why leaders fail in introducing values-based leadership? An elaboration of feasible steps, challenges, and suggestions for practitioners. *International Journal of Business and Management, 7*(9), 28–39.
Vithessonthi, C. (2009). Corporate ecological sustainability strategy decisions: The role of attitude towards sustainable development. *Journal of Organisational Transformation & Social Change, 6*(1), 49–64.
Vogel, D. J. (2005). Is there a market for virtue? The business case for corporate social responsibility. *California Management Review, 47*(4), 19–45.
Vormedal, I., & Ruud, A. (2009). Sustainability reporting in Norway–an assessment of performance in the context of legal demands and socio-political drivers. *Business Strategy and the Environment, 18*(4), 207–222.
Walden, W. D., & Schwartz, B. N. (1997). Environmental disclosures and public policy pressure. *Journal of Accounting and Public Policy, 16*(2), 125–154.
Wales, T. (2013). Organizational sustainability: What is it, and why does it matter. *Review of Enterprise and Management Studies, 1*(1), 38–49.
Weissman, S. H., & Sekutowski, J. C. (1991). Environmentally conscious manufacturing: A technology for the nineties. *AT&T Technical Journal, 70*(6), 23–30.

Welford, R. (1994). *Cases in environmental management and business strategy*. London: Rtman Publishing.

Welford, R. (1995). *Environmental strategy and sustainable development: The corporate challenge of the 21st century*. London: Routledge.

Welford, R. (1998). Corporate strategy, competitiveness and the environment. *Corporate Environmental Management, 1*, 13–34.

Welford, R. (2016a). *Corporate environmental management 1: Systems and strategies*. New York: Routledge.

Welford, R. (2016b). *Corporate environmental management 3: Towards sustainable development*. New York: Routledge.

Welford, R. (2016c). *Shared environmental values and organizational effectiveness. Corporate environmental management* 2 (pp. 120–136). New York: Routledge.

Wesselink, R., Blok, V., & Ringersma, J. (2017). Pro-environmental behaviour in the workplace and the role of managers and organisation. *Journal of Cleaner Production, 168*, 1679–1687.

Westerman, J. W., Nafees, L., & Westerman, J. (2021). When sustainability managers greenwash: SDG fit and effects on job performance and attitudes. *Sustainability, 13*(12), 65–69.

Wilbanks, T. (1994). Sustainable development in geographic perspective. *Annals of the Association of American Geographers, 84*(4), 541–556.

Zawawi, N. F. M., & Abd Wahab, S. (2019). Organizational sustainability: A redefinition? *Journal of Strategy and Management, 12*(3), 397–408.

Zhang, J., Ul-Durar, S., Akhtar, M. N., Zhang, Y., & Lu, L. (2021). How does responsible leadership affect employees' voluntary workplace green behaviors? A multilevel dual process model of voluntary workplace green behaviors. *Journal of Environmental Management, 296*, 113–205.

CHAPTER

8

Comparison of the SLM, SLS, and DLMS techniques in additive manufacture of AlSi10Mg alloys

Şenol Bayraktar and Cem Alparslan

Faculty of Engineering and Architecture, Department of Mechanical Engineering, Recep Tayyip Erdogan University, Rize, Turkey

8.1 History of additive manufacturing and its development

Demands for products to meet special needs are increasing day by day, spurred by the developing technology and increasing population. Accordingly, it is insufficient to find solutions for increasing demands by using conventional manufacturing methods. For this reason, additive manufacturing (AM), which is one of the technological methods that has developed in recent years, has contributed to the beginning of a new era in the manufacturing of industrial parts. AM is the production of a workpiece designed on a computer with three-dimensional (3D) software in layers using many different types of materials such as metal (Bian, Shamsaei, & John, 2017), composite (Mahamood, 2017), plastic (Tosto et al., 2020), and ceramic (Niu, Wu, Zhou, & Ma, 2014; Sames, List, Pannala, Dehoff, & Babu, 2016). The term *additive manufacturing* first entered the literature for model and prototype construction in the early 1980s (Wong & Hernandez, 2012). Chuck Hull manufactured 3D parts for the first time with a method of curing liquid polymer material (stereolithography) using a laser beam. The device was patented in 1984. Later, Hull cofounded the first 3D manufacturing company, 3D Systems, and the first commercial AM method emerged in 1987 (Wohlers & Gornet, 2014). Thanks to the studies that were carried out at

Innovation and Sustainable Manufacturing
DOI: https://doi.org/10.1016/B978-0-12-819513-0.00004-4

the University of Texas in 1986, the use of plastic powders with the laser sintering method was accepted as one of the AM methods. Fused deposition modeling (FDM) technology gained importance as a new method in the late 1980s. Scott Crump, who discovered the FDM method, founded a 3D printer company named Stratasys. The 3D printing technology became commercialized in the 1990s (Brooks, Kinsley, & Owens, 2014). A new technology for 3D printing was developed by the Massachusetts Institute of Technology in the mid-1990s. Multicolor production was developed after single-color production for additive production using this technology (Ishengoma & Mtaho, 2014). Although not all the studies were shared until the mid-2000s, 3D printers using the open-source extrusion method were launched in 2007. Thus the possibilities for developing and using 3D printers have increased (Jones et al., 2011). The development and use of AM methods after all these processes have increased. As a result, many companies have developed devices that are capable of manufacturing with various AM methods for both industry and professional users. Thanks to this technology, design flexibility, short manufacturing time, and the ability to create complex geometries have been achieved for parts that are difficult to manufacture. Workpieces are beginning to be efficiently manufactured in terms of cost and limited material waste by using AM (Fig. 8.1) (Gao et al., 2015).

8.1.1 Importance of additive manufacturing in industry

Thanks to the Industrial Revolution, autonomous and new industrial technology vehicles is being used in many fields instead of human

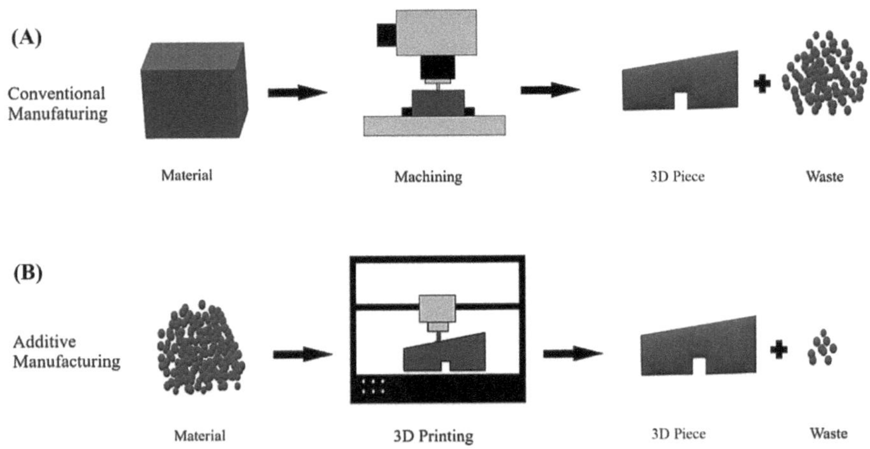

FIGURE 8.1 Manufacturing methods. (A) Conventional manufacturing method. (B) Additive manufacturing method.

production factors. Thus new technologies with high added value and competitive power have emerged in the manufacturing ecosystem. Significant changes have occurred in the workforce with all these developments. AM technologies, which are among these changes, have emerged as an alternative to conventional manufacturing methods. The desired workpiece can be manufactured in AM by adding layers on top of each other, unlike conventional substrate manufacturing methods. In this method the workpiece is drawn by using computer-aided design (CAD) programs and is saved as a file with the extension ".stl" (Wong & Hernandez, 2012). Then this workpiece model is loaded into appropriate 3D software. The model is divided into 2D layers within the software and sliced and made ready for manufacturing. The sliced model is sent to the 3D printer after the control process and manufacturing begins. The 3D printer performs layer-by-layer manufacturing using 2D geometric data (Kempen et al., 2011; Petrovic et al., 2011; Wong & Hernandez, 2012). At the last stage, the product that was manufactured with AM is separated from the support layer, and the manufacturing process is completed with different surface finishing processes, such as deburring (Fig. 8.2).

Certain manufacturing methods are taken into account in the design processes of workpieces used in mechanical systems. This situation bring about limitations in the design of the workpieces to be produced. AM is recognized as one of the modern manufacturing techniques. One of its most important advantages is that it allows unlimited design freedom.

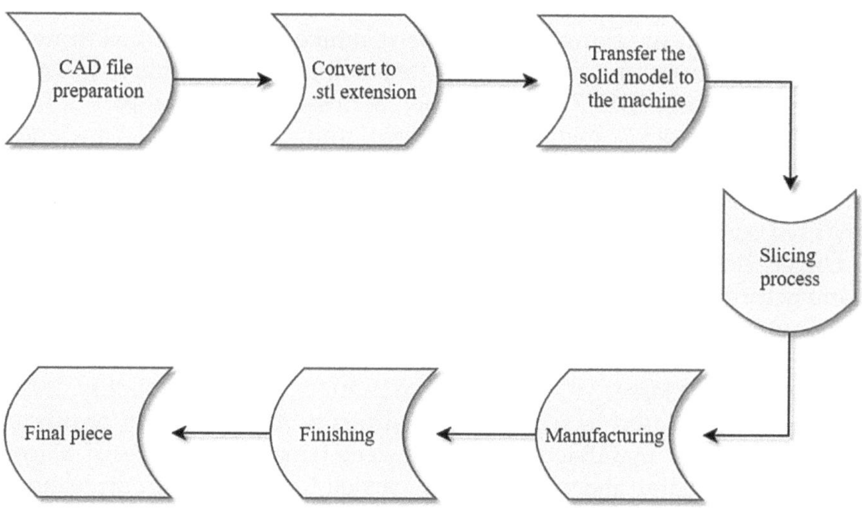

FIGURE 8.2 Stages of the additive manufacturing process.

Thus workpieces can be manufactured in desired geometries for consumer needs. Other advantages provided by AM include the following:

- Material production processes are reduced.
- The total cost is reduced.
- Product improvement accelerates.
- Consumer-specific products can be manufactured.
- Quick modeling of the products is made possible.
- Workpieces with very difficult and complex geometry can be manufactured.
- The product time in design is reduced.
- It provides harmony between design and production.
- The product can be functionally optimized.
- The number of manufacturing machines decreases.
- The delivery time is reduced.
- The stock requirement is low.
- There is no need for a mold.
- Consumption of materials is reduced.
- Logistics get easier.
- It enables rapid prototyping at low costs.

8.2 Materials used in the additive manufacturing method

The AM technique is based on the manufacturing of the base material layer by layer, unlike conventional subtractive manufacturing methods. Raw materials, which are in powder or wire form, can be selectively melted by the heat source and can be combined and solidified simultaneously in newly developing AM technologies (Emmelmann, Kranz, Herzog, & Wycisk, 2013; Carroll, Palmer, & Besse, 2015). Examples of these are thermoplastics, metals, ceramics, and biochemical materials. Thermoplastics are the most used materials in AM methods. Acrylonitrile butadiene styrene, polylactic acid, and polycarbonate materials are generally used as raw material called filament in wire form (Celik, Sener, Serin, & Unver, 2020; Zhou et al., 2021). Thanks to these materials, access to 3D manufacturing has become easier. Many materials, such as zirconia, alumina, and tricalcium phosphate, are preferred in AM. Especially today, powder glass and adhesive layers can be baked together to create new glass product classes (Grossin, et al., 2021; Saha & Mallik, 2021). Biochemicals are used to support bone structure for humans in healthcare applications. The manufacturing of hardened materials made of silicon, calcium phosphate, and zinc can be carried out as a support element (Garot, Bettega, & Picart, 2021; Sakthiabirami et al., 2021; Zafamris et al., 2021). In addition, many metals or metal alloys, such as stainless steel, titanium, and aluminum, are used in AM. The metal materials that are used are generally in powder form and are manufactured by water, gas, or plasma atomization. The most important advantage of this method is to ensure homogeneous spreading of the powder. The homogeneous powder structure affects the properties of the workpieces, such as density and porosity (Bourell et al., 2017). The metal powders that are used in AM must have certain physical and chemical properties in order to obtain a reproducible and reliable print quality. These properties are chemical composition, density, particle morphology, porosity, particle size distribution, fluidity, agglomeration, moisture, and tribocharge. To obtain high-quality and reliable workpieces using metal AM, each of these features must be optimized (Bian et al., 2017). Thanks to AM technology, these problems have been largely eliminated in the manufacturing of aluminum-based materials. Al-Si-Mg alloys are important alloys for the automotive industry, owing to their weldability, corrosion resistance, energy efficiency, and good resistance to cyclic loading. While the wear resistance of the workpieces is improved with Si additive, their strength and working resistance at high temperature can be improved with the addition of magnesium. In other words, the thermal and mechanical stability of the workpieces can be improved (Bayraktar & Demir, 2020; Kumar, Chakraborti, Lukas, Bodak, & Rokhlin, 2005). Manufacturing of these alloys is possible with AM technology (Guo, Cheng, & Liu, 2017). In addition, high processing speeds can be achieved, owing to the high thermal conductivity of aluminum-based parts. This provides an advantage for AM technology. The process parameters, microstructure, mechanical, and machinability properties of these alloys are generally examined in the workpieces manufactured with AM in the literature. Correct selection and optimization of these independent variable conditions are necessary for optimum product quality in the manufacturing industry.

8.3 Classification of additive manufacturing technologies

AM is classified according to energy source, process type, material, and material-manufacturing techniques. While the classification for the process type is laser and extrusion, powder, liquid and filament raw materials can be given as examples in terms of material (Gao et al., 2015; Gibson, Rosen, & Stucker, 2015). Classification for AM was performed by the American Society for Testing and Materials (ASTM) as shown in Table 8.1. AM techniques in this classification are categorized as binder jetting, directed energy deposition, material extrusion, material jetting, powder bed fusion, sheet lamination, and vat photopolymerization. Thus AM techniques can be used according to this standard in line with consumer needs (Tofail et al., 2018).

TABLE 8.1 Classification of additive manufacturing techniques for ASTM-52900:2015.

ASTM category	AM technology
Binder jetting	• 3D inkjet technology
Directed energy deposition	• Laser deposition • Laser engineered net shaping • Electron beam • Plasma arc melting
Material extrusion	• FDM/fused filament fabrication, fused layer modeling
Material jetting	• 3D inkjet technology • Direct Ink writing
Powder bed fusion	• Electron beam melting • Direct metal laser sintering • Selective laser sintering/melting
Sheet lamination	• Laminated object manufacturing • Ultrasound consolidation/ultrasound AM
Vat photopolymerization	• Stereolithography • Digital light processing

8.3.1 Selective laser sintering and selective laser melting

Selective laser sintering (SLS) technology is a powder bed AM technique. The basic logic of this method is based on the principle of combining metal powders by means of a laser beam. SLS was developed by Dr. Carl Deckard and his academic advisor Dr. Joe Beaman at the University of Texas in the mid-1980s (Deckard, 1989). Powders in this technique are spread in layers and sintered to manufacture the workpiece. The powder is evenly distributed and spread by a flattening roller to obtain a smooth structure in the manufacturing of the workpiece. Then a CO_2 laser beam scans the cross section of the part by directing the powder on the manufacturing plane to the desired layers. Heat occurs at the points where the beam hits the surface during scanning. The powder material with the effect of this temperature is partially sintered and fuses with other powder grains in contact. A platform is placed under the machine table to create the other layer after scanning moves down as much as the layer thickness. Powder is spread on the previous scanned layer by using rollers as much as the layer thickness, and the laser scanning process continues. All these process steps continue until the workpiece manufacturing is completed (Fig. 8.3) (Vaezi, Seitz, & Yang, 2013). After the AM process, the workpiece is removed from the machine table, and free dust is cleaned from it with a brush or vacuum system (Partee, Hollister, & Das, 2006).

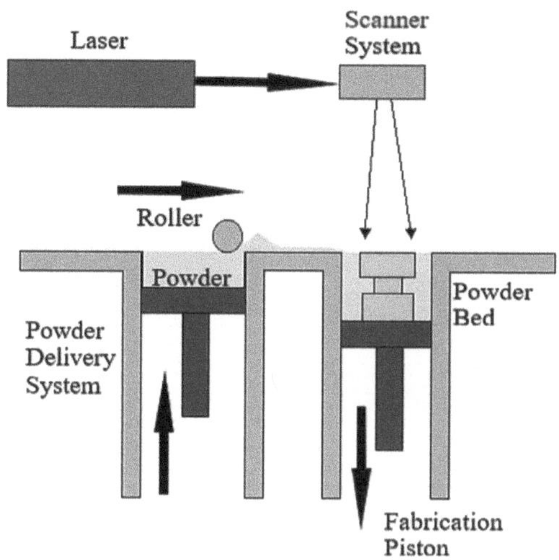

FIGURE 8.3 Schematic representation of selective laser sintering technology.

SLS prints are extremely detailed and often mechanically superior to workpieces manufactured by using other technologies. SLS is used mostly for fully functional prototypes in advanced materials. A major advantage of SLS is that support structures are not needed. Continuous powder spreading provides all the necessary support for the workpiece. Therefore SLS can be used to create freeform geometries that are impossible to manufacture by any other method. Because the bond strength between layers is excellent in SLS, it has excellent tensile strength and modulus relative to the casted material. However, it is more fragile. SLS technology has many advantages, such as high part accuracy in manufacturing, applicability for different materials, no need for part support during manufacturing, usability for workpieces with complex geometry, easy prototyping, and a fast manufacturing process. Plastics, such as polyamide, polystyrene, and carbon fiber, and different metal powder mixtures, such as aluminum, stainless steel, chrome, and nickel, can be used as raw materials in the SLS method (Vaezi et al., 2013; Yakout, Elbestawi, & Veldhuis, 2018). The selective laser melting (SLM) manufacturing technique is similar to SLS. While sintering is used in SLS, a melting technique is used in SLM. Complex 3D models can be manufactured in layers in a short time. Metal, polymer, ceramic, and composite-based powders are used as raw materials in SLM technique (Kumar, 2003). A schematic of SLM technology is given in Fig. 8.4. Here, powder is spread on the machine table as much as the layer thickness. The powder is combined by penetration with a laser beam at a temperature below the melting point of the material. This process is repeated for each layer until the workpiece is manufactured (Kempen

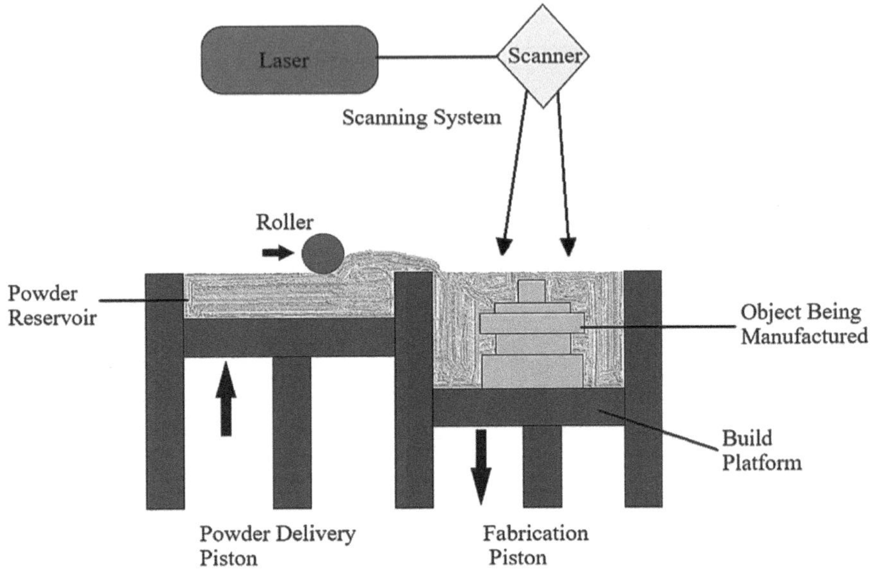

FIGURE 8.4 Schematic representation of selective laser melting technology.

et al., 2011). The most important parameter in this method is optimum energy density. The most important parameters of energy density are laser power and scanning speed. Manufacturing of workpieces is completed by melting the powder grains in layers with the energy density created by choosing the optimum scanning speed and laser power according to the material type (Shirazi et al., 2015).

The purpose of using high energy is to obtain the ideal density for the combination of powder particles. The peak power of the Nd:YAG laser that is used in SLM can be up to 3 kW. However, about 50 W of power is used in practice. The microstructure and thermal residual stresses in the manufactured workpieces can be reduced by using thermal annealing. This can improve the surface quality of some workpieces. Thus there is no need for a final process after the manufacturing of the materials (Bourell, Evan, Chen, & Barrow, 2004). The process is carried out in an argon atmosphere to minimize oxygen and nitrogen uptake. The thickness of each layer varies between 0.05 and 0.1 mm in SLM systems. When the process parameters are not controlled in this method, high internal stresses and low surface quality occur in the manufactured workpieces. In particular, high temperature changes during the process can cause this situation (Tolochko et al., 2004). The SLM technique is an excellent technology for the rapid manufacturing of specific products. Products with a very complex structure in the fields of machinery, medicine, and art can be manufactured easily. It is a suitable method for parts and designs that require constant revision. This method is more costly than other methods for parts that do not change their design and require large quantities of manufacturing. In addition, the manufacturing of workpieces with very large dimensions is quite difficult in using this method (Louvis, Fox, & Sutcliffe, 2011; Yap et al., 2015). Workpieces can be manufactured by using aluminum, cobalt, chrome, stainless steel, Inconel, and titanium powders with this method. In some of the current studies of the SLM technique in the literature, Bharath et al. investigated the mechanical behavior of selective laser melted AlSi10Mg and die cast A360 alloys. Tensile strength, yield strength, and percent failure strain were found to be significantly higher in selective laser melted material than in A360 alloy. It was observed that the anisotropic properties of parts that were manufactured by SLM were minimal, and highly compatible results were obtained as a result of the absence of large differences in stress-strain curves (Bharath, Shamanth, & Hemanth, 2021). Zaretsky et al. comparatively analyzed the dynamic responses of AlSi10Mg alloy produced by SLM and by a sand-casting technique. They determined that the dynamic responses in the alloy that was manufactured with SLM were independent of the process direction. The yield and tensile strength of selective laser melted material compared to the casted alloy were found to be two and four times higher, respectively. In addition, it was found that this difference was due to the fact that the materials that were manufactured with SLM had a very high homogeneous structure (Zaretsky, Stern, & Frage, 2017). Nurel et al. tested the dynamic properties of AlSi10Mg alloy manufactured using SLM with the split Hopkinson pressure bar test. Radial-type cracks were observed on the fracture surfaces of the as-built alloy, while ductile deformation was observed on the fracture surfaces of T5 heat-treated workpieces (Nurel, Nahmany, Frage, Stern, & Sadot, 2018). Maconachie et al. researched the effect of different build orientation on the quasistatic and dynamic responses of SLMed AlSi10Mg alloy. While little difference was observed between the build orientations in terms of strength, it was found that elements that were loaded perpendicular to the build direction were more ductile than other build orientations. Build orientation was found to be have significant effect on fracture surface morphology, owing to the fracture paths following melt pool boundaries. Fracture surfaces were observed along the melt pool boundaries due to the strong, brittle α-Al and eutectic Si formed by the melting of the previous layers during the SLM (Fig. 8.5) (Maconachie et al., 2020).

Yan et al. comparatively investigated the effect of microstructure on the mechanical properties of AlSi10Mg alloy manufactured by SLM and casting techniques. A supersaturated Si network structure was observed throughout the α-Al phase in the selective laser melted alloy. It was determined that brittle β and π phases were formed, depending on thermal properties in as-cast and selective laser melted alloys, respectively.

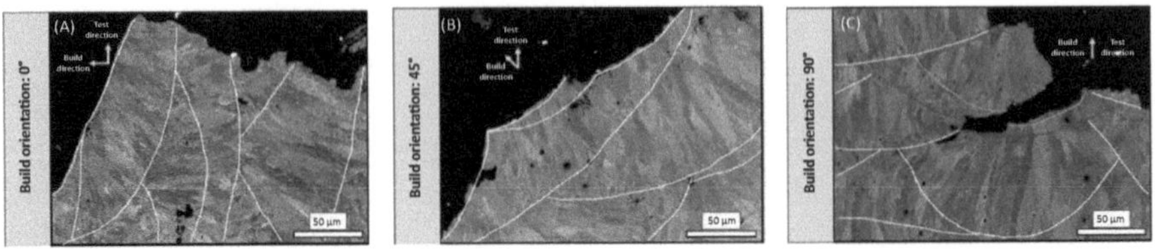

FIGURE 8.5 Cross-sectional images of fracture surface of AlSi10Mg alloy in different build orientations. (A) 0 degrees. (B) 45 degrees. (C) 90 degrees. *From Tofail, S.A.M., Koumoulos, E.P., Koumoulos, Bandyopadhyay, A., Bose, S., Odoghue, L., Charitidis, C. (2018). Additive manufacturing: Scientific and technological challenges, market uptake and opportunities,* Materials Today, 21*(1), 22–37. Reproduced with permission from Elsevier.*

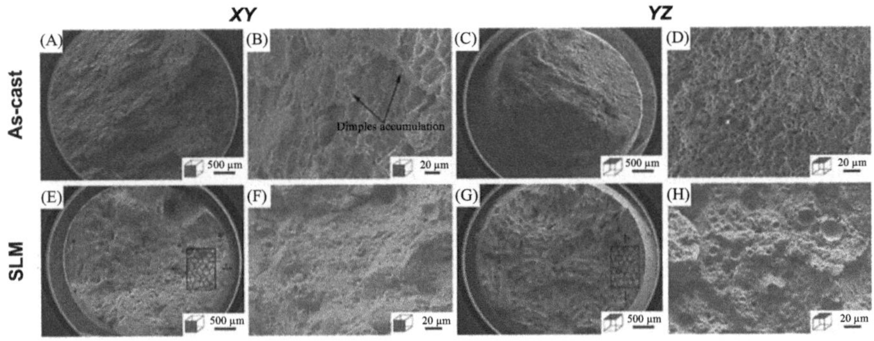

FIGURE 8.6 Fracture surface images for (A, B, C, D) as-cast and (E, F, G, H) selective laser melted AlSi10Mg built along the XY direction (A, B, E, F) and the YZ direction (C, D, G, H). *From Yan, Q., Song, B., & Shi, Y. (2020). Comparative study of performance comparison of AlSi10Mg alloy prepared by selective laser melting and casting,* Journal of Materials Science & Technology, 41, *199–208. Reproduced with permission from Elsevier.*

Higher tensile strength was measured in selective laser melted alloy compared to as-cast alloy due to grain refining, while elongation to fracture was measured lower. This indicated that the selective laser melted alloy was more brittle (Fig. 8.6). Unmelted powder particles, oxide inclusion, and pores in selective laser melted alloy were reported to had a negative effect on fatigue life (Yan, Song, & Shi, 2020).

Park et al. investigated the effect of direct aging without solution heat treatment on the microstructure and tensile properties of SLMed AlSi10Mg alloy. Conventional T6 heat treatment was applied for comparison. The researchers determined that the tensile strength results at room temperature were higher (310 MPa) in the direct aged alloy than in the as-built and T6 heat-treated alloys. The elongation to break of the direct aged alloy was observed to be 3% lower than that of the as-built alloy, while it was higher than that of the T6 heat-treated alloy. A fine dimple formation was observed in the as-built alloy in the direction of the fracture surface images, and the cellular structure was effective in increasing the alloy strength (Fig. 8.7) (Park, Baek, Hyer, Sohn, & Lee, 2021).

Li et al. researched the effect of construction angles (0 degrees, 15 degrees, 30 degrees, 45 degrees, 60 degrees, 75 degrees, and 90 degrees) on the microstructure and mechanical properties of SLMed AlSi10Mg alloy. They found that the 45-degree sample had the highest hardness value and the 60 degrees sample had a good combination of strength and plasticity, with a tensile strength of 463.54 MPa, a yield strength of 283.37 MPa, and an elongation of 9.25%. They showed that the grain structure of the alloy gradually changed from columnar grains to equiaxed grains in the vertical plane with increasing construction angle (Fig. 8.8) (Li et al., 2021).

Wang et al. examined the effect of different heat treatment temperatures (270°C, 300°C, and 330°C) and building orientations (0 degrees, 45 degrees, and 90 degrees) on the anisotropic tensile properties of the SLMed AlSi10Mg alloy. They observed that the white cell structure in the microstructure was Si eutectic and that the Si gridlike structure was distributed around α-Al dendrites (Fig. 8.9). It was shown that the grain size did not increase significantly with heat treatment using electron backscatter diffraction. The orientation angle size was observed to be inversely proportional to the plasticity. The researchers stated that this

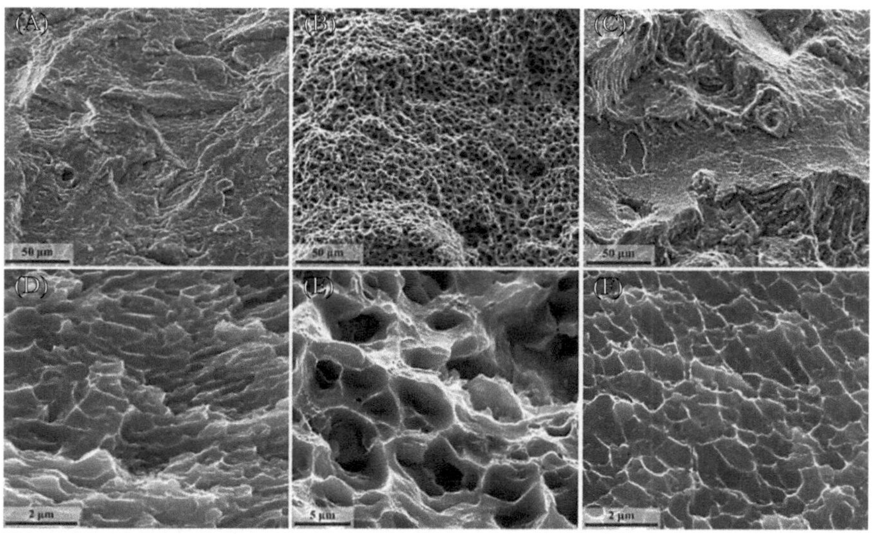

FIGURE 8.7 Fracture surface images. (A, D) As-built. (B, E) T6 heat-treated. (C, F) Direct aged alloy. *From Park, T.H., Baek, M.S., Hyer, H., Sohn, Y., Lee, K.A. (2021). Effect of direct aging on the microstructure and tensile properties of AlSi10Mg alloy manufactured by selective laser melting process,* Materials Characterization, 176, *111113. Reproduced with permission from Elsevier.*

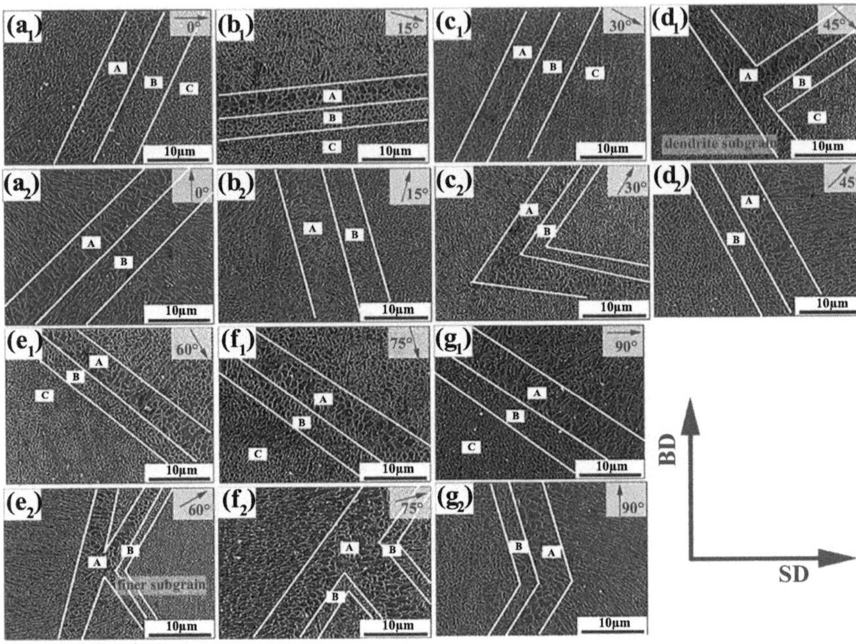

FIGURE 8.8 SLMed AlSi10Mg samples for different construction angles. (a_1−g_1) Horizontal plane. (a_2−g_2) Vertical plane. *From Li, X., Yi, D., Wu, X., Zhang, J., Yang, X., Zhao, Z., Liu, Y. (2021). Effect of construction angles on microstructure and mechanical properties of AlSi10Mg alloy fabricated by selective laser melting,* Journal of Alloys and Compounds, 881, *160459. Reproduced with permission from Elsevier.*

was due to different degrees of dislocation pile-up in vertical and horizontal directions with the effect of eutectic Si (Wang et al., 2022).

8.3.2 Direct metal laser sintering

Direct metal laser sintering (DMLS) technology is one of the most widely used powder metal AM methods. Aluminum, cobalt chromium, nickel, and titanium-based alloys and refractive metal, stainless, and tool steels can be manufactured by using this method. This technique is based on the principle of sintering the metal powder mixture with high-energy laser beams without melting it. The sintering process takes place by combining the metal powders by means of diffusion mechanisms to melt the base material. The working schematic of the DMLS technique is similar to that of SLM. However, powders are sintered at the molecular level in the SLM method, unlike the melting process. Thus the formation of porosity in the workpieces is minimized. In addition, this situation allows the manufacturing of alloys containing materials with

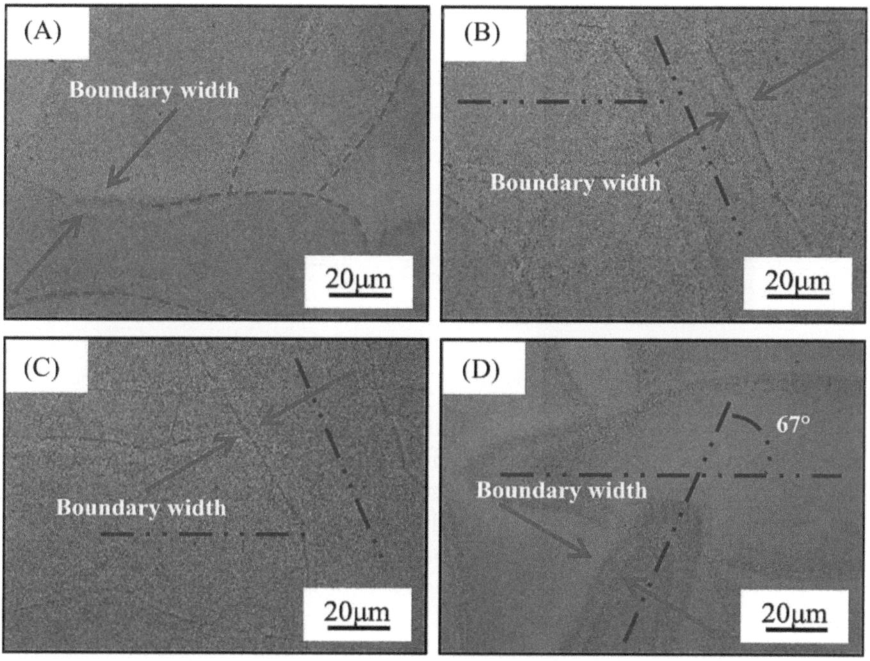

FIGURE 8.9 Optical microscope images of the tensile specimen. (A) 90 degrees, heat treated, 270°C, 2 h. (B) 90 degrees, heat treated, 300°C, 2 h. (C) 90 degrees, heat treated, 330°C, 2 h. (D) 90 degrees, as-built. *From Wang, C.G., Zhu, J.X., Wang, G.W., Qin, Y., Sun, M.Y., Yang, J.L., (2022). Huang, S.K. Effect of building orientation and heat treatment on the anisotropic tensile properties of AlSi10Mg fabricated by selective laser melting,* Journal of Alloys and Compounds, *895, 162665. Reproduced with permission from Elsevier.*

different melting points. SLS is used in processes that are applied to different materials (plastic, glass, and ceramic), while DMLS is used for metal alloys. A schematic of DMLS technology is given in Fig. 8.10. The manufacturing process in this technique is started by sending the product file designed using the CAD program to the appropriate machine. Metal powder during the manufacturing process is spread on the machine platform at the desired layer thickness using a roller or a spreading tool. The powder spreading thickness (layer thickness), melt penetration, and powder dimensions are determined according to the manufacturing parameters. The powder that has been spread on the platform is sintered with high-energy fiber/CO_2 laser beams in line with the design data. The powder spreading and sintering processes are repeated until the desired product geometry is completed. After all layers have been sintered sequentially, the final product emerges.

It is possible to print directly with metal and powder mixtures in DMLS technology, and workpieces with complex geometries can be easily manufactured. Materials with different structures and unsintered metal powders

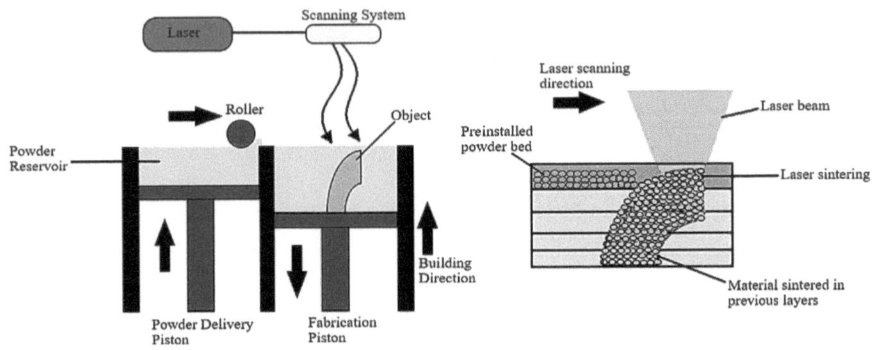

FIGURE 8.10 Schematic representation of direct metal laser sintering technology.

can be reused and enable the manufacturing of durable and functional workpieces. DMLS is a unique method for the manufacturing of workpieces with complex and special geometries. However, the manufacturing time, manufacturing stages, and cost higher than conventional manufacturing methods limit the usage areas of this technology. This technology is often preferred for the manufacturing of prototypes or small numbers of workpieces (Kayacan & Yılmaz, 2019). In some studies on direct metal laser sintered AlSi10Mg alloy in the literature, Hadadzadeh et al. investigated the effect of building direction on the transition from columnar to equiaxed structure during the manufacturing of AlSi10Mg alloy with DMLS. They showed that both dendritic and granular structures changed from columnar to equiaxed with the change in building direction from vertical to horizontal. They observed that 75% of the grains in the vertical sample were columnar, and when changed horizontally, 49% of the grains became columnar. They found that very finely dispersed and compatible Si precipitates were formed in the α-Al dendrites of the vertical samples. They also showed that the dislocation density decreased and the size of Si precipitates increased in the horizontal sample (Fig. 8.11) (Hadadzadeh, Amirkhiz, Li, & Mohammadi, 2018).

Hadadzadeh et al. researched the strength behavior of direct metal laser sintered AlSi10Mg alloy manufactured in different building directions under $1400\ s^{-1}$ strain rate and dynamic loading using a split Hopkinson pressure bar. It was observed that fine and compatible Si precipitates were formed in the vertical sample (Fig. 8.12), while coarse and semicompatible Si precipitates were formed in the horizontal sample. The dislocation density increased three times with the change of building direction from horizontal to vertical. The dynamic loading behaviors in vertical and horizontal samples were almost similar (Hadadzadeh, Amirkhiz, Odeshi, & Mohammadi, 2018).

Cabrini et al. examined the effect of heat treatment on the corrosion resistance of direct metal laser sintered AlSi10Mg alloy. Local corrosion

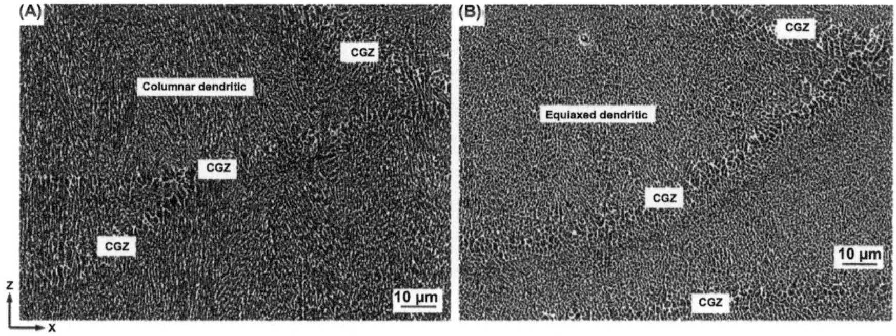

FIGURE 8.11 Direct metal laser sintered AlSi10Mg alloy. (A) Vertical specimen. (B) Horizontal specimens. *From Hadadzadeh A., Amirkhiz, B.S., Odeshi, A., Mohammadi, M. (2018). Dynamic loading of direct metal laser sintered AlSi10Mg alloy: Strengthening behavior in different building directions,* Materials and Design, 159, *201–211. Reproduced with permission from Elsevier.*

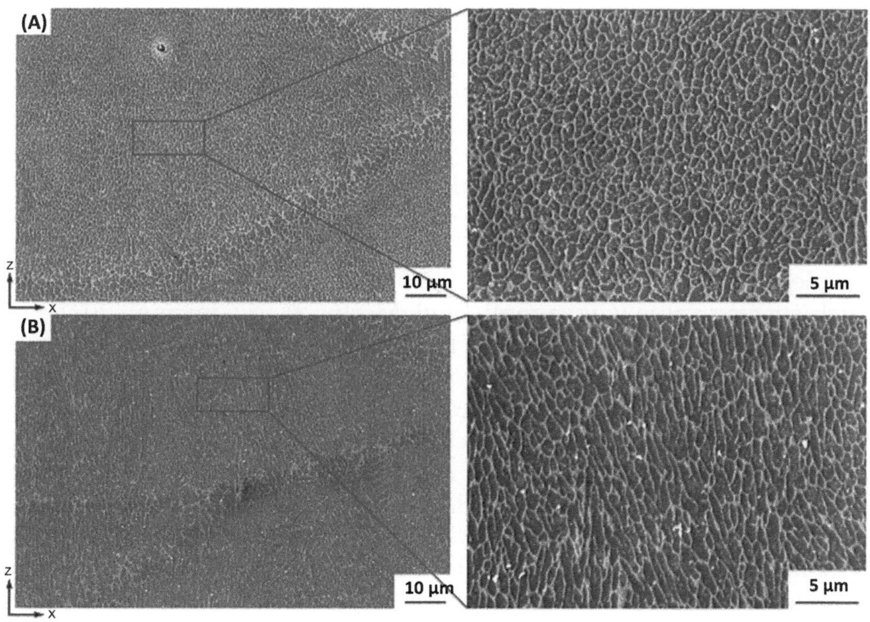

FIGURE 8.12 Microstructure of direct metal laser sintered AlSi10Mg alloy. (A) Horizontal specimen. (B) Vertical specimen. (*Right side*) Details for the interior structure of melt pools. *From Hadadzadeh A., Amirkhiz, B.S., Odeshi, A., Mohammadi, M. (2018). Dynamic loading of direct metal laser sintered AlSi10Mg alloy: Strengthening behavior in different building directions,* Materials and Design, 159, *201–211. Reproduced with permission from Elsevier.*

and selective penetrating attack were observed at the melt pool boundary in untreated or only stress relieved samples. While no selective attack was observed after annealing at high temperatures, a galvanic linkage was observed with coarse Si particles that were extensively precipitated. It was revealed that the annealing process performed at 823 K for 4 hours contributes to the incorporation of Si particles (Cabrini et al., 2016). Sahoo researched on the modeling of the temperature profile formed during the manufacturing of direct metal laser sintered AlSi10Mg alloy with ANSYS. While it was observed that the laser energy input had a significant effect on the temperature increase in the build layer during the manufacturing of the alloy, it was determined that it was inversely proportional to the scanning speed. It was revealed that the melt pool dimensions increase with the increase of laser energy density and the porosity played a critical role in determining the thermal distribution throughout the powder bed. In addition, porosity and thermal conductivity factors were found to be related to each other for metals (Sahoo, 2021). Giovagnoli et al. investigated the effect of process parameters on the porosity and mechanical properties of direct metal laser sintered AlSi10Mg alloy using energy density approach. It was determined that the limited variation of the process parameters had a significant effect on the porosity of the alloy but had a negligible effect on the microstructural properties. It was observed that the high elongation to break value was measured as a result of the increase in porosity, and this was due to the micro-scale fracture behavior. While weak fracture paths were formed in dense samples (Fig. 8.13A and B), keyhole-shaped porosities were observed in porous samples (Fig. 8.13D). The samples exhibited better tensile properties with the decrease of the hatch-spacing value (Giovagnoli, Silvi, Merlin, & Di Giovanni, 2021).

Hadadzadeh et al. investigated the strength properties of direct metal laser sintered AlSi10Mg alloy produced by using recycled powder at certain process parameters (laser power: 370 W, scan speed: 1300 mm/s, hatch distance: 190 μm, layer thickness: 30 μm). The microstructure of the alloy was observed to consist of primary α-Al bounded by eutectic Si walls in both directions (Fig. 8.14). Direct metal laser sintered alloy exhibited higher semistatic uniaxial tensile strength compared to cast alloy in both vertical and horizontal build directions. The strengthening of the direct metal laser sintered alloy manufactured in both directions was detailed with reference to the Orowan mechanism, the Hall-petch effect, and the dislocation hardening. It was revealed that dislocations interacted with eutectic Si walls, and Si precipitated (Hadadzadeh, Baxter, Amirkhiz, & Mohammadi, 2018).

Ghasri-Khouzani et al. researched the hardness and microstructural properties of AlSi10Mg alloy manufactured by using direct metal laser

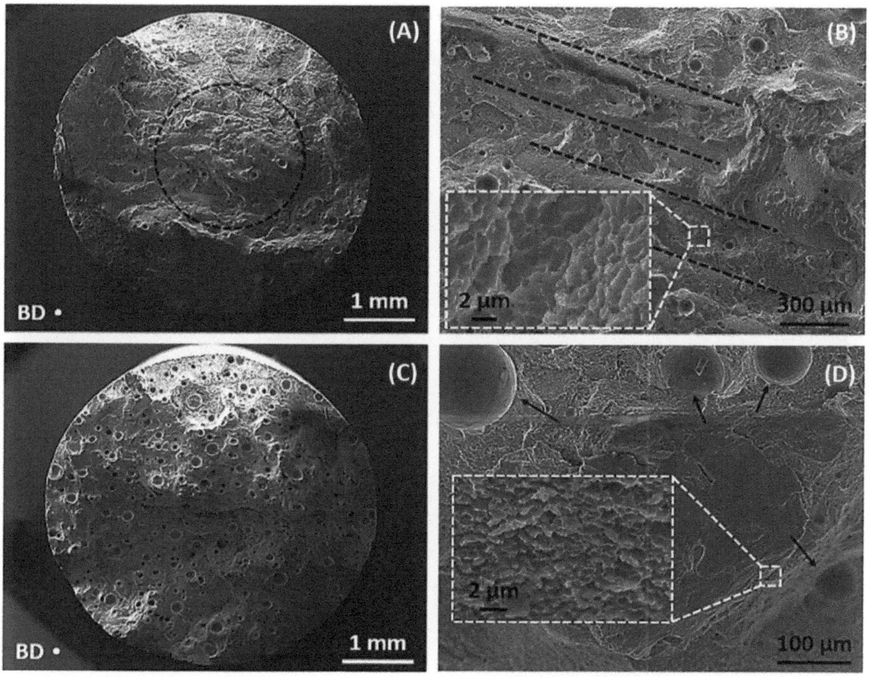

FIGURE 8.13 Fracture surface images for alloy samples at different energy densities. (A, B) Low energy density. (C, D) High energy density. *From Giovagnoli, M., Silvi, G., Merlin, M., Di Giovanni, M.T. (2021). Optimisation of process parameters for an additively manufactured AlSi10Mg alloy: Limitations of the energy density-based approach on porosity and mechanical properties estimation,* Materials Science and Engineering: A 802, *140613. Reproduced with permission from Elsevier.*

sintered technique for different planes. It was determined that the morphology of many cells changed as XY (horizontal) equiaxed the columnar structure on the XZ (vertical) plane, and a fine cellular dendritic network structure was observed (Fig. 8.15). It was found that the morphology of cells that were subjected to solidification was strongly dependent on the morphology of the grains that formed them. According to the X-ray diffraction analysis it was observed that similar phase aggregates were formed in both planes, and the microhardness difference between the planes was insignificant (3%) (Ghasri-Khouzani et al., 2019).

8.4 Concluding remarks

Today, many manufacturing technologies have been developed with similar approaches based on AM technology. Methods such as stereolithography, FDM, 3D printing, SLS, SLM, DMLS, and electron beam

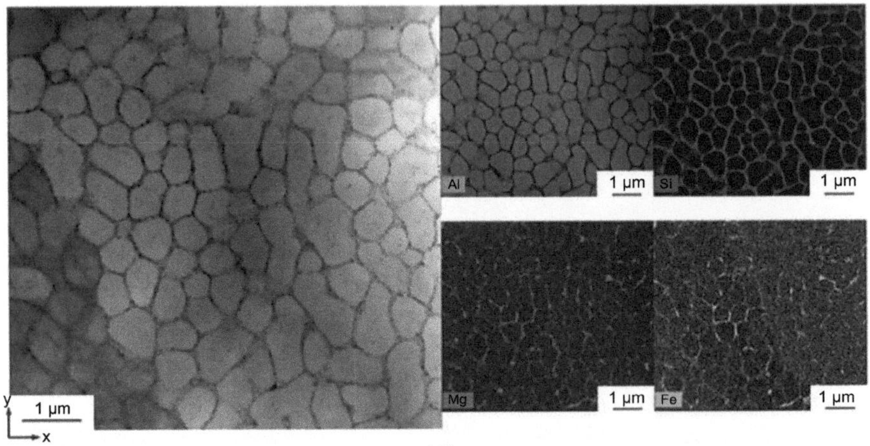

(A)

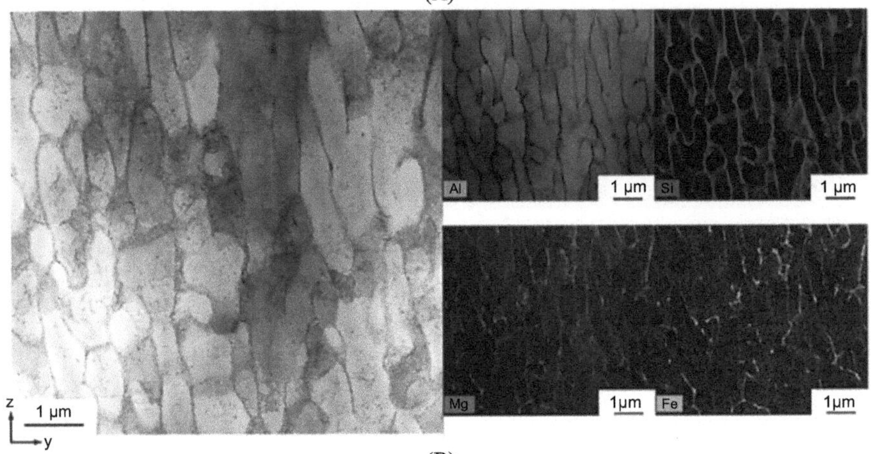

(B)

FIGURE 8.14 Scanning transmission electron microscope bright field microstructure for (A) vertical and (B) horizontal samples. *From Hadadzadeh A., Amirkhiz, B.S., Odeshi, A., Mohammadi, M. (2018). Dynamic loading of direct metal laser sintered AlSi10Mg alloy: Strengthening behavior in different building directions,* Materials and Design, 159, *201–211. Reproduced with permission from Elsevier.*

melting are the most widely used among them. The difficulties that are encountered in conventional manufacturing methods can be minimized by using these methods. In particular, they provide design freedom in workpieces that require molds and fixtures and allow the manufacturing of workpiece materials that contain different elements. As a result of these features, they can quickly meet consumer needs in the automotive, energy, space, aviation, and medical industries. Many material groups based on aluminum, titanium, and nickel can be manufactured under

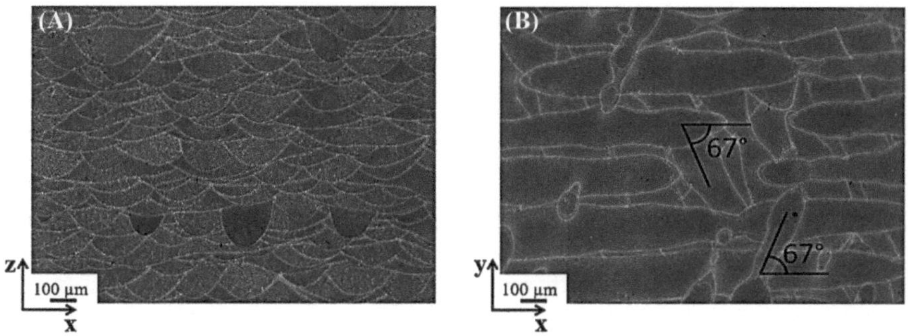

FIGURE 8.15 Optical microscope images of direct metal laser sintered AlSi10Mg alloy at different cross sections. (A) Vertical (front surface). (B) Horizontal (top surface). *From Ghasri-Khouzani, M., Peng, H., Attardo, R., Ostiguy, P., Neidig, J., Billo, R., Shankar, M.R. (2019). Comparing microstructure and hardness of direct metal laser sintered AlSi10Mg alloy between different planes,* Journal of Manufacturing Processes, *37, 274–280. Reproduced with permission from Elsevier.*

suitable conditions with these methods. They are ideal methods for manufacturing prototypes and limited numbers of workpieces. Thus it is possible to control the initial investment costs. Powder materials are spread on a machine table layer by layer in the powder bed fusion method used in the AM technique. Each layer is processed in a controlled manner by melting only the necessary areas with a laser. SLS, which is one of the powder bed fusion methods, uses the laser source only to form the bond of particles with each other in melting the powder. In SLS, more vibration is applied to the building platform to ensure homogeneous distribution of powder in each region of the layer compared to other methods. The surface roughness of the workpieces that are manufactured with SLS is higher than in the SLM and DMLS methods. The structures of SLS and DMLS are similar. However, SLS can be used with different metals and nonmetals, while DMLS is used for metals only. Although SLM is a faster process than the SLS method, the use of inert gases such as nitrogen and argon in SLM increases energy costs compared to the SLS method. Powder particles in SLM are not only fused to each other, but melted in a homogeneous form. Therefore the strength of the workpieces that are manufactured with SLM technology is generally higher than that in other techniques. While metal powders at a single melting temperature are completely melted in SLM, the powders in DMLS consist of materials with variable melting points and fuse at high temperatures at the molecular level. In other words, the SLM method can manufacture workpieces containing a single metal element, and metal alloys containing more than one element can be manufactured by using DMLS. Generally, workpieces that are manufactured with the SLM process have high surface quality, while workpieces with higher dimensional stability can be manufactured with DMLS. The most important advantage of DMLS over other techniques is the manufacturing of workpieces with minimal residual stress and structural defects. Therefore DMLS is preferred in the manufacturing of workpieces that can work in environments with high stresses, such as in the aerospace and automotive sectors. In this chapter, SLM, SLS, and DMLS processes used in AM methods were detailed based on AlSi10Mg alloy. Independent variables such as build direction, platform temperature, scanning distance, energy density, and powder layer thickness of the parts manufactured in general were examined in the studies found in the literature using these processes. The effects of independent variable parameters on hardness, wear resistance, microstructure properties, fatigue resistance, yield strength, surface quality, and corrosion resistance were also revealed. It was determined that the thermophysical properties of the materials were effective on the local melting of the alloying elements between the layers with the heat source. While the DMLS or SLM method was preferred in the manufacturing of AlSi10Mg alloy with AM, it was observed that SLS method was not preferred much because more efficient results can be obtained for prototyping metal workpieces and low volume manufacturing by using the DMLS and SLM processes. As a result of the manufacturing of AlSi10Mg alloy by the DMLS technique, it was observed that less porous structures were formed in the internal structure in comparison to the SLM method. There is still a need to develop new techniques for improving the surface quality so that the produced workpieces can be used directly in mechanical systems. In addition, the need for lower-cost and high-stability AM systems in terms of mass production is increasing. It is not enough to optimize a large number of independent variables with processes such as Taguchi and RSM using today's AM techniques. More efficient results can be obtained for optimization with the use of multicriteria systems such as TOPSIS, PROMETHEE, ELECTRE, AHP, WASPAS, COPRAS, and PSO. Thus time and cost savings can be achieved for minimum surface roughness and high mechanical properties in workpieces that can be manufactured with the AM process.

References

Bayraktar, Ş., & Demir, O. (2020). Processing of T6 heat-treated Al-12Si-0.6 Mg alloy. *Materials and Manufacturing Processes, 35*(3), 354–362.

Bharath, C., Shamanth, V., & Hemanth, K. (2021). Studies on mechanical behaviour of AlSi10Mg alloy produced by selective laser melting and A360 alloy by die casting. *Materials Today: Proceedings, 45*, 78–81.

Bian, L., Shamsaei, N., & John, M. U. (2017). *Laser-based additive manufacturing of metal parts: Modeling, optimization, and control of mechanical properties*. CRC Press-Taylor & Francis.

Bourell, D., Kruth, J. P., Leu, M., Levy, G., Rosen, D., Beese, A. M., & Clare, A. (2017). Materials for additive manufacturing. *CIRP Annals, 66*(2), 659–681.

Bourell, D. L., Evan, R. S., Chen, S. W., & Barrow, S. L. (2004). Rapid manufacturing of functional parts using İnfiltration post-processing. In: *Proceedings of the fourth laser assisted net shape engineering, LANE 1* (pp. 93–104).

Brooks, G., Kinsley, K., & Owens, T. (2014). 3D printing as a consumer technology business model. *International Journal of Management & Information Systems, 18*(4), 271. Available from https://doi.org/10.19030/ijmis.v18i4.8819.

Cabrini, M., Lorenzi, S., Pastore, T., Pellegrini, S., Ambrosio, E. P., Calignano, F., ... Fino, P. (2016). Effect of heat treatment on corrosion resistance of DMLS AlSi10Mg alloy. *Electrochimica Acta, 206*, 346–355.

Carroll, B. E., Palmer, T. A., & Besse, A. M. (2015). Anisotropic tensile behavior of Ti-6Al-4V components fabricated with directed energy deposition additive manufacturing. *Acta Materialia, 87*, 309–320. Available from https://doi.org/10.1016/j.actamat.2014.12.054.

Celik, B., Sener, B., Serin, G., & Unver, H. O. (2020). Ergiyik filament fabrikasyonu, 3B yazıcılar için kompozit filament ekstrüder makinesi geliştirilmesi. *Makina Tasarım ve İmalat Dergisi, 17*(2), 65–75.

Deckard, C. (1989). *Method and apparatus for producing parts by selective sintering*, US Patent 4,863,538, filed 17 October 1986, published 5 September 1989.

Emmelmann, C., Kranz, J., Herzog, D., & Wycisk, E. (2013). Laser additive manufacturing of metals. *Laser technology in biomimetics* (pp. 143–161). Heidelberg: Springer. Available from http://doi.org/10.1007/978-3-642-41341-4_6.

Gao, W., Zhang, Y., Ramanujan, D., Ramani, K., Chen, Y., Williams, C. B., ... Zavattieri, P. D. (2015). The status, challenges, and future of additive manufacturing in engineering. *Computer-Aided Design, 69*, 65–89. Available from https://doi.org/10.1016/j.cad.2015.04.001.

Garot, C., Bettega, G., & Picart, C. (2021). Additive manufacturing of material Scaffolds for bone regeneration: Toward application in the clinics. *Advanced Functional Materials, 31*(5), 2006967.

Ghasri-Khouzani, M., Peng, H., Attardo, R., Ostiguy, P., Neidig, J., Billo, R., ... Shankar, M. R. (2019). Comparing microstructure and hardness of direct metal laser sintered AlSi10Mg alloy between different planes. *Journal of Manufacturing Processes, 37*, 274–280.

Gibson, I., Rosen, D., & Stucker, B. (2015). Directed energy deposition processes. *Additive Manufacturing Technologies, 2*, 245–268.

Giovagnoli, M., Silvi, G., Merlin, M., & Di Giovanni, M. T. (2021). Optimisation of process parameters for an additively manufactured AlSi10Mg alloy: Limitations of the energy density-based approach on porosity and mechanical properties estimation. *Materials Science and Engineering: A, 802*, 140613.

Grossin, D., Montón, A., Navarrete-Segado, P., Özmen, E., Urruth, G., Maury, F., Bertrand, G. (2021). A review of additive manufacturing of ceramics by powder bed selective laser processing (sintering/melting): Calcium phosphate, silicon carbide, zirconia, alumina, and their composites. *Open Ceramics*, 100073.

Guo, X., Cheng, G., & Liu, W. K. (2017). *Report of the workshop predictive theoretical, computational and experimental approaches for additive manufacturing (WAM 2016)*. Switzerland: Springer.

Hadadzadeh, A., Amirkhiz, B. S., Odeshi, A., & Mohammadi, M. (2018). Dynamic loading of direct metal laser sintered AlSi10Mg alloy: Strengthening behavior in different building directions. *Materials and Design, 159*, 201–211.

Hadadzadeh, A., Amirkhiz, B. S., Li, J., & Mohammadi, M. (2018). Columnar to equiaxed transition during direct metal laser sintering of AlSi10Mg alloy: Effect of building direction. *Additive Manufacturing, 23*, 121–131.

Hadadzadeh, A., Baxter, C., Amirkhiz, B. S., & Mohammadi, M. (2018). Strengthening mechanisms in direct metal laser sintered AlSi10Mg: Comparison between virgin and recycled powders. *Additive Manufacturing, 23*, 108–120.

Ishengoma, F. R., & Mtaho, A. B. (2014). 3D printing: Developing countries perspectives. *International Journal of Computer Applications- arXiv preprint, 104*(11), 30–34. Available from https://doi.org/10.5120/18249-9329.

Jones, R., Haufe, P., Sells, E., Iravani, P., Olliver, V., Palmer, C., & Bowyer, A. (2011). RepRap–the replicating rapid prototyper. *Robotica, 29*(1), 177–191. Available from https://doi.org/10.1017/S026357471000069X.

Kayacan, M. Y., & Yılmaz, N. (2019). DMLS additive manufacturing process and cost model development. *Journal of Polytechnic, 22*(3), 763–770.

Kempen, K., Thijs, L., Yasa, E., Badrossamay, M., Verheecke, W., & Kruth, J. P. (2011). Process optimization and microstructural analysis for selective laser melting of AlSi10Mg. In: *Solid freeform fabrication symposium proceedings* (pp. 484–495). Texas. Available from https://doi.org/10.26153/tsw/15310.

Kumar, K. H., Chakraborti, N., Lukas, H., Bodak, O., & Rokhlin, L. (2005). *Light metal ternary systems: Phase diagrams, crystallographic and thermodynamic data, Al-Mg-Si (Aluminium–Magnesium–Silicon).* Berlin: Springer Verlag. Available from http://doi.org/10.1007/10915998_18.

Kumar, S. (2003). Selective laser sintering: A qualitative and objective approach. *JOM, 55*(10), 43–47. Available from https://doi.org/10.1007/s11837-003-0175-y.

Li, X., Yi, D., Wu, X., Zhang, J., Yang, X., Zhao, Z., & Liu, Y. (2021). Effect of construction angles on microstructure and mechanical properties of AlSi10Mg alloy fabricated by selective laser melting. *Journal of Alloys and Compounds, 881*, 160459.

Louvis, E., Fox, P., & Sutcliffe, C. J. (2011). Selective laser melting of aluminium components. *Journal of Materials Processing Technology, 211*, 275–284.

Maconachie, T., Leary, M., Zhang, J., Medvedev, A., Sarker, A., Ruan, D., ... Brandt, M. (2020). Effect of build orientation on the quasi-static and dynamic response of SLM AlSi10Mg. *Materials Science and Engineering A, 788*, 139445.

Mahamood, R. M. (2017). *Laser metal deposition process of metals, alloys, and composite materials.* Cham: Springer International Publishing. Available from https://doi.org/10.1007/978-3-319-64985-6.

Niu, F., Wu, D., Zhou, S., & Ma, G. (2014). Power prediction for laser engineered net shaping of Al_2O_3 ceramic parts. *Journal of the European Ceramic Society, 34*, 3811–3817. Available from https://doi.org/10.1016/j.jeurceramsoc.2014.06.023.

Nurel, B., Nahmany, M., Frage, N., Stern, A., & Sadot, O. (2018). Split Hopkinson pressure bar tests for investigating dynamic properties of additively manufactured AlSi10Mg alloy by selective laser melting. *Additive Manufacturing, 22*, 823–833.

Park, T. H., Baek, M. S., Hyer, H., Sohn, Y., & Lee, K. A. (2021). Effect of direct aging on the microstructure and tensile properties of AlSi10Mg alloy manufactured by selective laser melting process. *Materials Characterization, 176*, 111113.

Partee, B., Hollister, S. J., & Das, S. (2006). Selective laser sintering process optimization for layered manufacturing of CAPA® 6501 polycaprolactone bone tissue engineering scaffolds. *Journal of Manufacturing Science and Engineering (ASME), 128*, 531–540.

Petrovic, V., Gonzalez, J. V. H., Ferrando, O. J., Gordillo, J. D., Puchades, J. R. B., & Grinan, L. P. (2011). Additive layered manufacturing: Sectors of industrial application shown through case studies. *International Journal of Production Research, 49*, 1061–1079. Available from https://doi.org/10.1080/00207540903479786.

Saha, M., & Mallik, M. (2021). Additive manufacturing of ceramics and cermets: Present status and future perspectives. *Sādhanā, 46*(3), 1–35.

Sahoo, S. (2021). Direct metal laser sintering of AlSi10Mg alloy parts: Modeling of temperature profile. *Materials Today: Proceedings, 35*, 118–123.

Sakthiabirami, K., Kang, J. H., Jang, J. G., Soundharrajan, V., Lim, H. P., Yun, K. D., & Park, S. W. (2021). Hybrid porous zirconia scaffolds fabricated using additive manufacturing for bone tissue engineering applications. *Materials Science and Engineering: C, 123*, 111950.

Sames, W. J., List, F. A., Pannala, S., Dehoff, R. R., & Babu, S. S. (2016). The metallurgy and processing science of metal additive manufacturing. *International Materials Reviews, 61*(5), 1–46. Available from https://doi.org/10.1080/09506608.2015.1116649.

Shirazi, S. F. S., Gharehkhani, S., Mehrali, M., Yarmand, H., Metselaar, H. S. C., Kadri, N. A., & Osman, N. A. A. (2015). A review on powder-based additive manufacturing for tissue engineering: Selective laser sintering and inkjet 3D printing. *Science and Technology of Advanced Materials, 16*, 1–21.

Tofail, S. A. M., Koumoulos, E. P., Bandyopadhyay, A., Bose, S., Odoghue, L., & Charitidis, C. (2018). Additive manufacturing: Scientific and technological challenges, market uptake and opportunities. *Materials Today, 21*(1), 22–37.

Tolochko, N. C., Mozzharov, S. E., Yadroitsev, I. A., Laoui, T., Froyen, L., Titov, V. I., & Ignatiev, M. B. (2004). Balling processes during selective laser treatment of powders. *Rapid Prototyping Journal, 10*(2), 78–87. Available from https://doi.org/10.1108/13552540410526953.

Tosto, C., Latteri, A., Pergolizzi, E., Giordano, D., Abramo, G., Catenaro, R., ... Cicala, G. (2020). Additive manufacturing of plastics: An efficient approach for composite tooling, In *Macromolecular Symposia, 389*(1), 1900069. Available from https://doi.org/10.1002/masy.201900069.

Vaezi, M., Seitz, H., & Yang, S. (2013). A review on 3D micro-additive manufacturing technologies. *International Journal of Advanced Manufacturing Technology, 67*(5–8), 1721–1754.

Wang, C. G., Zhu, J. X., Wang, G. W., Qin, Y., Sun, M. Y., Yang, J. L., ... Huang, S. K. (2022). Effect of building orientation and heat treatment on the anisotropic tensile properties of AlSi10Mg fabricated by selective laser melting. *Journal of Alloys and Compounds, 895*, 162665.

Wohlers, T., & Gornet, T. (2014). History of additive manufacturing. *Wohlers report, 24* (2014), 118.

Wong, K. V., & Hernandez, A. (2012). A review of additive manufacturing. *ISRN Mechanical Engineering, 4*, 1–10. Available from https://doi.org/10.5402/2012/208760.

Yakout, M., Elbestawi, M. A., & Veldhuis, C. (2018). A review of metal additive manufacturing technologies. *Solid State Phenomena, 278*, 1–14.

Yan, Q., Song, B., & Shi, Y. (2020). Comparative study of performance comparison of AlSi10Mg alloy prepared by selective laser melting and casting. *Journal of Materials Science & Technology, 41*, 199–208.

Yap, C. Y., Chua, C. K., Dong, Z. L., Liu, Z. H., Zhang, D. Q., Loh, L. E., & Sing, S. L. (2015). Review of selective laser melting: Materials and applications. *Applied Physics Reviews, 2*, 1–21.

Zafamris, K., Brasinika, D., Karatza, A., Koumoulos, E., Karoussis, I. K., Kyriakidou, K., & Charitidis, C. A. (2021). Additive manufacturing of hydroxyapatite–chitosan–genipin composite scaffolds for bone tissue engineering applications. *Materials Science and Engineering: C, 119*, 111639.

Zaretsky, E., Stern, A., & Frage, N. (2017). Dynamic response of AlSi10Mg alloy fabricated by selective laser melting. *Materials Science and Engineering A, 688*, 364–370.

Zhou, X., Deng, J., Fang, C., Lam, W., Song, Y., Zhang, Z., & Li, Y. (2021). Additive manufacturing of CNTs/PLA composites and the correlation between microstructure and functional properties. *Journal of Materials Science & Technology, 60*, 27–34.

Index

Note: Page numbers followed by "*f*" and "*t*" refer to figures and tables, respectively.

D

E

Kolibri 360 GmbH
Ferdinand-Jühlke-Straße 7
99095 Erfurt, Deutschland
produktsicherheit@kolibri360.de